Evaluating Research Efficiency
in the U.S. Environmental Protection Agency

Committee on Evaluating the Efficiency of Research and Development
Programs at the U.S. Environmental Protection Agency

Committee on Science, Engineering, and Public Policy

Policy and Global Affairs

Board on Environmental Studies and Toxicology

Division on Earth and Life Studies

NATIONAL RESEARCH COUNCIL
OF THE NATIONAL ACADEMIES

THE NATIONAL ACADEMIES PRESS
Washington, D.C.
www.nap.edu

THE NATIONAL ACADEMIES PRESS   **500 Fifth Street, NW**   **Washington, DC 20001**

NOTICE: The project that is the subject of this report was approved by the Governing Board of the National Research Council, whose members are drawn from the councils of the National Academy of Sciences, the National Academy of Engineering, and the Institute of Medicine. The members of the committee responsible for the report were chosen for their special competences and with regard for appropriate balance.

This project was supported by Contract 68-C-03-081 between the National Academy of Sciences and the U.S. Environmental Protection Agency. Any opinions, findings, conclusions, or recommendations expressed in this publication are those of the authors and do not necessarily reflect the view of the organizations or agencies that provided support for this project.

International Standard Book Number-13 978-0-309-11982-5
International Standard Book Number-10 0-309-11982-0

Additional copies of this report are available from

The National Academies Press
500 Fifth Street, NW
Box 285
Washington, DC 20055

800-624-6242
202-334-3313 (in the Washington metropolitan area)
http://www.nap.edu

THE NATIONAL ACADEMIES
Advisers to the Nation on Science, Engineering, and Medicine

The **National Academy of Sciences** is a private, nonprofit, self-perpetuating society of distinguished scholars engaged in scientific and engineering research, dedicated to the furtherance of science and technology and to their use for the general welfare. Upon the authority of the charter granted to it by the Congress in 1863, the Academy has a mandate that requires it to advise the federal government on scientific and technical matters. Dr. Ralph J. Cicerone is president of the National Academy of Sciences.

The **National Academy of Engineering** was established in 1964, under the charter of the National Academy of Sciences, as a parallel organization of outstanding engineers. It is autonomous in its administration and in the selection of its members, sharing with the National Academy of Sciences the responsibility for advising the federal government. The National Academy of Engineering also sponsors engineering programs aimed at meeting national needs, encourages education and research, and recognizes the superior achievements of engineers. Dr. Charles M. Vest is president of the National Academy of Engineering.

The **Institute of Medicine** was established in 1970 by the National Academy of Sciences to secure the services of eminent members of appropriate professions in the examination of policy matters pertaining to the health of the public. The Institute acts under the responsibility given to the National Academy of Sciences by its congressional charter to be an adviser to the federal government and, upon its own initiative, to identify issues of medical care, research, and education. Dr. Harvey V. Fineberg is president of the Institute of Medicine.

The **National Research Council** was organized by the National Academy of Sciences in 1916 to associate the broad community of science and technology with the Academy's purposes of furthering knowledge and advising the federal government. Functioning in accordance with general policies determined by the Academy, the Council has become the principal operating agency of both the National Academy of Sciences and the National Academy of Engineering in providing services to the government, the public, and the scientific and engineering communities. The Council is administered jointly by both Academies and the Institute of Medicine. Dr. Ralph J. Cicerone and Dr. Charles M. Vest are chair and vice chair, respectively, of the National Research Council.

www.national-academies.org

COMMITTEE ON SCIENCE, ENGINEERING, AND PUBLIC POLICY

Members

GEORGE WHITESIDES (*Chair*), Woodford L. and Ann A. Flowers University Professor, Harvard University, Boston, MA

CLAUDE R. CANIZARES, Vice President for Research, Massachusetts Institute of Technology, Cambridge

RALPH J. CICERONE (Ex officio), President, National Academy of Sciences, Washington, DC

EDWARD F. CRAWLEY, Executive Director, CMI, and Professor, Massachusetts Institute of Technology, Cambridge

RUTH A. DAVID, President and Chief Executive Officer, Analytic Services, Inc., Arlington, VA

HAILE T. DEBAS, Executive Director, UCSF Global Health Sciences, Maurice Galante Distinguished Professor of Surgery, San Francisco, CA

HARVEY FINEBERG (Ex officio), President, Institute of Medicine, Washington, DC

JACQUES S. GANSLER, Vice President for Research, University of Maryland, College Park

ELSA M. GARMIRE, Professor, Dartmouth College, Hanover, NH

M. R. C. GREENWOOD (Ex officio), Professor of Nutrition and Internal Medicine, University of California, Davis

W. CARL LINEBERGER, Professor of Chemistry, University of Colorado, Boulder

C. DAN MOTE, JR. (Ex officio), President and Glenn Martin Institute Professor of Engineering, University of Maryland, College Park

ROBERT M. NEREM, Parker H. Petit Professor and Director, Institute for Bioengineering and Bioscience, Georgia Institute of Technology, Atlanta

LAWRENCE T. PAPAY, Retired, Sector Vice President for Integrated Solutions, Science Applications International Corporation, La Jolla, CA

ANNE C. PETERSEN, Professor of Psychology, Stanford University, Stanford, CA

SUSAN C. SCRIMSHAW, President, Simmons College, Boston, MA

WILLIAM J. SPENCER, Chairman Emeritus, SEMATECH, Austin, TX

LYDIA THOMAS (Ex officio), Retired, Mitretek Systems, Inc., Falls Church, VA

CHARLES M. VEST (Ex officio), President, National Academy of Engineering, Washington, DC

NANCY S. WEXLER, Higgins Professor of Neuropsychology, Columbia University, New York, NY

MARY LOU ZOBACK, Vice President Earthquake Risk Applications, Risk Management Solutions, Inc., Newark, CA

Staff

RICHARD BISSELL, Executive Director
DEBORAH STINE, Associate Director (up to August 2007)
MARION RAMSEY, Administrative Coordinator
NEERAJ P. GORKHALY, Senior Program Assistant

An Assessment of the National Science Foundation's Science and Technology Centers Program (1996)

Allocating Federal Funds for Science and Technology (1995)

Reshaping the Graduate Education of Scientists and Engineers (1995)

On Being a Scientist: Responsible Conduct in Research (1995)

Major Award Decisionmaking at the National Science Foundation (1994)

Science, Technology, and the Federal Government: National Goals for a New Era (1993)

Responsible Science Volume 2: Background Papers and Resource Documents (1993)

Responsible Science Volume 1: Ensuring the Integrity of the Research Process (1992)

Policy Implications of Greenhouse Warming: Mitigation, Adaptation, and the Science Base (1991)

*Copies of these reports may be ordered from the National Academies Press
(800) 624-6242 or (202) 334-3313
www.nap.edu*

Ecological Indicators for the Nation (2000)
Waste Incineration and Public Health (2000)
Hormonally Active Agents in the Environment (1999)
Research Priorities for Airborne Particulate Matter (four volumes, 1998-2004)
The National Research Council's Committee on Toxicology: The First 50
        Years (1997)
Carcinogens and Anticarcinogens in the Human Diet (1996)
Upstream: Salmon and Society in the Pacific Northwest (1996)
Science and the Endangered Species Act (1995)
Wetlands: Characteristics and Boundaries (1995)
Biologic Markers (five volumes, 1989-1995)
Science and Judgment in Risk Assessment (1994)
Pesticides in the Diets of Infants and Children (1993)
Dolphins and the Tuna Industry (1992)
Science and the National Parks (1992)
Human Exposure Assessment for Airborne Pollutants (1991)
Rethinking the Ozone Problem in Urban and Regional Air Pollution (1991)
Decline of the Sea Turtles (1990)

Copies of these reports may be ordered from the National Academies Press
(800) 624-6242 or (202) 334-3313
www.nap.edu

Preface

In an effort to ensure the wise use of taxpayers' money, the federal government has undertaken major initiatives to evaluate the performance and results of federally funded programs, including research and development (R&D) programs. The initiatives include the Government Performance and Results Act of 1993 and the Office of Management and Budget (OMB) Program Assessment Rating Tool (PART), developed in 2002. The latter was designed in the context of "performance budgeting" and "performance measurement" and focused on evaluating the efficiency of programs.

Evaluation of R&D programs has proved challenging for federal agencies, including the Environmental Protection Agency (EPA), a mission-oriented agency with a substantial research component. All agencies have experienced difficulties in complying with the PART requirements to measure the efficiency of their research, to use outcome-based metrics in doing so, and to achieve and quantitate annual efficiency improvements.

In 2006, EPA asked the National Academies for assistance in developing better assessment tools to comply with PART, with emphasis on efficiency. The Academies' Committee on Science, Engineering, and Public Policy (COSEPUP) and the National Research Council Board on Environmental Studies and Toxicology (BEST) convened the Committee on Evaluating the Efficiency of Research and Development Programs at the U.S. Environmental Protection Agency.

In this report, the committee addresses the efficiency measures now used for federally funded R&D programs and evaluates whether they are sufficient and are based on outcomes, not only inputs and outputs. The committee provides principles that it hopes will guide the development of efficiency measures for federally funded R&D programs and makes recommendations about efficiency measures for EPA's basic and applied R&D programs and about OMB's review process.

The committee gratefully acknowledges the following for making presentations to the committee: Diana Espinosa, Daren Wong, Brian Kleinman, and Kevin Neyland, of OMB; Marcus Peacock, Hugh Tilson, Phillip Juengst, Lori Kowalski, of EPA; Bill Valdez, Darrell Beschen, and Brian Card, of the Department of Energy; Pat Tsuchitani, of the National Science Foundation; Deborah Duran, of the National Institutes of Health; Julie Pollitt, of the National Aeronautics and Space Administration; Raymond Sinclair, of the National Institute for Occupational Safety and Health; George Daston, of the Procter & Gamble Company; Joseph Kenney and Bernice Rogowitz, of IBM Global Business Services; James Bus, of Dow Chemical Company; and Patrick Atkins, retired from Alcoa.

The committee is grateful for the assistance of the National Research Council staff in preparing this report. Staff members who contributed to the effort are Richard Bissell, project director from August 2007 and director of COSEPUP; Deborah Stine, project director (up to August 2007); Eileen Abt, senior program officer; James Reisa, director of BEST; Alan Anderson, consultant science writer; Jennifer Saunders, associate program officer; Rae Benedict, Mirzayan Science & Technology Policy Fellow; Norman Grossblatt, senior editor; Mirsada Karalic-Loncarevic, manager of the Toxicology Information Center; and Neeraj Gorkhaly and Morgan Motto, senior project assistants.

I especially thank my colleagues on the committee for their engagements in the contentious issues underlying our charge and for what we believe are constructive principles and feasible recommendations that have emerged from our deliberations and the iterative development of this report.

Gilbert S. Omenn, *Chair*
Committee on Evaluating the Efficiency of
Research and Development Programs at the
U.S. Environmental Protection Agency

Acknowledgment of Review Participants

This report has been reviewed in draft form by persons chosen for their diverse perspectives and technical expertise in accordance with procedures approved by the National Research Council's Report Review Committee. The purpose of the independent review is to provide candid and critical comments that will assist the institution in making its published report as sound as possible and to ensure that the report meets institutional standards of objectivity, evidence, and responsiveness to the study charge. The review comments and draft manuscript remain confidential to protect the integrity of the deliberative process. We thank the following for their review of the report: Irwin Feller, Pennsylvania State University; Haren S. Gandhi, Ford Motor Company; Bernard D. Goldstein, University of Pittsburgh; Victoria F. Haynes, Research Triangle Institute; Martha A. Krebs, California Energy Commission; James H. Johnson, Jr., Howard University; Genevieve M. Matanoski, Johns Hopkins University; Shelley H. Metzenbaum, University of Maryland School of Public Policy; and David Trinkle, University of California, Berkeley.

Although the reviewers listed above have provided many constructive comments and suggestions, they were not asked to endorse the conclusions or recommendations, nor did they see the final draft of the report before its release. The review of the report was overseen by John F. Ahearne, Sigma Xi, The Scientific Research Society, and Alan Schriesheim, Argonne National Laboratory. Appointed by the National Research Council, they were responsible for making certain that an independent examination of the report was carried out in accordance with institutional procedures and that all review comments were carefully considered. Responsibility for the final content of this report rests entirely with the author committee and the institution.

Contents

BOXES, FIGURES, AND TABLES

BOXES

FIGURES

TABLES

Evaluating Research Efficiency
in the U.S. Environmental Protection Agency

Summary

The federal government has long sought effective tools to evaluate the performance and results of publicly funded programs, including research and development (R&D) programs, to ensure the wise use of taxpayers' money. To that end, Congress passed the Government Performance and Results Act in 1993, and the Office of Management and Budget (OMB) designed the Program Assessment Rating Tool (PART) in 2002.

Evaluation of R&D programs has proved to be challenging for federal agencies. In particular, they have experienced difficulties in complying with the PART requirements to measure the *efficiency* of their research, to use *outcome-based* metrics in doing so, and to achieve *annual* efficiency improvements.

In 2006, the U.S. Environmental Protection Agency (EPA) asked the National Academies for independent assistance in developing better assessment tools to comply with PART. The Academies' Committee on Science, Engineering, and Public Policy (COSEPUP) and the National Research Council (NRC) Board on Environmental Studies and Toxicology (BEST) oversaw the appointment of the Committee on Evaluating the Efficiency of Research and Development Programs at the U.S. Environmental Protection Agency and charged it to answer the following questions:

- What efficiency measures are currently used for EPA R&D programs and other federally funded R&D programs?
- Are these efficiency measures sufficient? Are they outcome-based?
- What principles should guide the development of efficiency measures for federally funded R&D programs?
- What efficiency measures should be used for EPA's basic and applied R&D programs?

Through a series of information-gathering steps, including discussions with OMB and EPA and a public workshop attended by representatives of research-intensive agencies[1] and industries, the committee evaluated how EPA and other agencies were attempting to comply with PART. The committee focused its deliberations on several fundamental issues posed by the charge questions, including

1. How—and why—should research be evaluated in terms of efficiency?
2. What is a "sufficient" measure of efficiency?
3. What measures of efficiency are "outcome-based," and should they be?

In its discussion the committee uses the terms inputs, outputs, and outcomes as defined by OMB, except as modified and discussed below:

- *Inputs* are agency resources—such as funding, facilities, and human capital—that support research.
- *Outputs* are activities or accomplishments delivered by research programs, such as research findings, papers published, exposure methods developed and validated, and research facilities built or upgraded.
- *Outcomes* are the benefits resulting from a research program, which can be short-term, such as an improved body of knowledge or a comprehensive science assessment, or long-term, such as lives saved or enhancement of air quality, that may be based on research activities or informed by research but that require additional activities by many others. The committee distinguishes these two types of outcomes using the terms, *intermediate outcomes* and *ultimate* or *end outcomes.*[2]

QUESTION 1

With respect to the question, "How—and why—should research be evaluated in terms of efficiency?", the committee suggests that some of the frustration expressed by federal research-intensive agencies in complying with PART derives from confusion over the concept of "efficiency." From its review of the OMB PART guidance and efficiency measures used by EPA and other federal agencies, the committee concludes that two conceptually different kinds of efficiency are integral to the execution and evaluation of R&D programs. The committee distinguished between *investment efficiency* and *process efficiency.*

Investment efficiency focuses on portfolio management, including the need to identify the most promising lines of research for achieving desired outcomes.

---

[1] The term *research-intensive* is used to describe agencies for which research is an essential even if not necessarily dominant aspect of the mission. For example, research is important at EPA but is not its primary function, as is the case for the National Institutes of Health and the National Science Foundation.

[2] The committee acknowledges that the NRC Committee for the Review of NIOSH Research Program has used the term end outcomes.

It is best evaluated by assessing the program's research activities, from planning to funding to midcourse adjustments, in the framework of its strategic planning architecture. Investment efficiency concerns three questions: are the right investments being made, is the research being performed at a high level of quality, and are timely and effective adjustments made in the multi-year course of the work to reflect new scientific information, new methods, and altered priorities? Because these questions cannot be addressed quantitatively, they require judgment based on experience and should be addressed through expert review.

Process efficiency involves inputs and outputs. Its evaluation asks how well research processes are managed. It monitors activities, such as publications, grants reviewed and awarded, and laboratory analyses conducted whose results can be anticipated and can be tracked quantitatively against established benchmarks by using such units as dollars and hours. Examples may include time required to conduct site assessments, average cost per measurement or analysis, and what percentage of external grants are evaluated by peer review within a given period.

Whereas both kinds of efficiency are addressed in concept in the PART questions, only the questions regarding process efficiency are labeled by PART guidance as efficiency per se. Operationally, though, OMB seeks to address these process efficiency questions using measures of outcomes.

QUESTION 2

In exploring the question, "What is a 'sufficient' measure of efficiency?", the committee assembled a list of relevant issues and examined a number of metrics proposed or used by federal agencies. It found that none of those metrics was capable of evaluating investment efficiency, and that many of the ones that were appropriate for evaluating process efficiency were not sufficient. Many of the process-efficiency metrics proposed by agencies other than EPA have been accepted by OMB, but several similar metrics proposed by EPA have not been accepted. Metrics that typically have been proposed or used by the federal agencies address only a small piece of a research program, and none attempts a comprehensive program evaluation.

QUESTION 3

In addressing the question, "What measures of efficiency are "outcome-based," and should they be?", the committee distinguished "ultimate outcomes," such as lives saved or clean air, from "intermediate outcomes," such as timely submission of comprehensive science assessments for scheduled regulatory reviews. While intermediate outcomes can be useful metrics, the committee found that *ultimate-outcome-based metrics* cannot be used to evaluate the efficiency of research for three reasons:

- Ultimate outcomes usually cannot be predicted or known in advance.
- Ultimate outcomes may occur long after research is completed.
- Ultimate outcomes usually depend on actions taken by others.

The PART *guidance* urges agencies to develop outcome-based efficiency metrics,[3] even though the PART *questions* do not specifically refer to such metrics and no agencies have been able to develop them for research programs.

PART also requires that assessments be made annually. That is difficult for research managers whose long-term projects may show results only after several years of work.

FINDINGS

The committee identified the following findings:

1. The key to research efficiency is good planning and implementation. EPA and its ORD have a sound strategic planning architecture that provides a multi-year basis for the annual assessment of progress and milestones for evaluating research programs, including their efficiency.

2. All the metrics examined by the committee that have been proposed by or accepted by OMB to evaluate the efficiency of federal research programs have been based on the inputs and outputs of research-management processes, not on their outcomes.

3. Ultimate-outcome-based efficiency metrics are neither achievable nor valid for this purpose.

4. EPA's difficulties in complying with the PART questions about efficiency (questions 3.4 and 4.3[4]) have grown out of inappropriate OMB requirements for outcome-based efficiency metrics.

5. An "ineffective"[5] PART rating of a research program can have serious adverse consequences for the program or the agency.

---

[3]For example, the PART guidance (p. 10) states, "Outcome efficiency measures are generally considered the best type of efficiency measure for assessing the program overall."

[4]Question 3.4: "Does the program have procedures (e.g. competitive sourcing/cost comparisons, IT improvements, appropriate incentives) to measure and achieve efficiencies and cost effectiveness in program execution?" Question 4.3: "Does the program demonstrate improved efficiencies or cost effectiveness in achieving program goals each year?"

[5]OMB PART Web site (http://www.expectmore.gov) states that "programs receiving the Ineffective rating are not using tax dollars effectively. Ineffective programs have been unable to achieve results due to a lack of clarity regarding the program's purpose or goals, poor management, or some other significant weakness. Ineffective programs are categorized as Not Performing."

6. Among the metrics proposed to measure process efficiency, several can be recommended for wider use by agencies.

7. The most effective mechanism for evaluating the investment efficiency of R&D programs is an expert-review panel, as recommended in earlier COSEPUP and BEST reports. Expert-review panels are much broader than scientific peer-review panels.

RECOMMENDATIONS

Recommendation 1

To comply with questions 3.4 and 4.3 of PART, EPA and other agencies should only apply quantitative efficiency metrics to measure the *process efficiency* of research programs. Process efficiency can be measured in terms of inputs, outputs, and some intermediate outcomes; it does not require ultimate outcomes.

For compliance with PART, evaluation of the efficiency of a research program should not be based on ultimate outcomes, for the reasons listed above under Question 3. Although PART guidance encourages the use of outcome-based metrics, the guidance also describes the difficulty of applying them. As stated earlier, the committee has concluded that, for most research programs, ultimate-outcome-based efficiency measures are *neither achievable nor valid*.

Given the inability to evaluate the efficiency of research on the basis of ultimate outcomes, the committee recommends that OMB and other oversight bodies focus on evaluating the *process efficiency* of research, including core research or basic research—how program managers exercise skill and prudence in using and conserving resources. For evaluating process efficiency, quantitative methods can be used by expert-review panels and others to track and review the use of resources in light of goals embedded in strategic and multi-year plans. Earned Value Management (EVM) is a quantitative tool that can track aspects of research programs against milestones.[6]

Moreover, to facilitate the evaluation process, the committee recommends modifying OMB's framework of results to include the category of *intermediate outcomes*, as distinguished from *ultimate outcomes*. Intermediate outcomes include such results as an improved body of knowledge available for decision-making, comprehensive science assessments, and the dissemination of newly developed tools and models. Those results, which might be visualized as intermediate between outputs and ultimate outcomes, might enhance the evaluation

---

[6]EVM measures the degree to which research outputs conform to scheduled costs along a time line. It is used by agencies and other organizations in many management settings, such as construction projects and facilities operations, where the outcome (such as a new laboratory or optimal use of facilities) is well known in advance and progress can be plotted against milestones.

process by adding individual trackable items and a larger body of knowledge for decision-making.

Recommendation 2

EPA and other agencies should use expert-review panels to evaluate the *investment efficiency* of research programs. The process should begin by evaluating the relevance, quality, and performance[7] of the research.

Investment efficiency is used in this report to indicate whether an agency is "doing the right research and doing it well." The term is meant as a gauge of portfolio management to measure whether a program manager is investing in research that is relevant to the agency's mission and long-term plans and is being performed at a high level of quality. Evaluating quality and relevance requires expert judgment based on experience; no quantitative measures can fully capture these key items. The best mechanism for measuring investment efficiency is the expert-review panel. Investment efficiency may also include studies that guide the next set of research projects or stepwise development of analytic tools or other products.

EPA should continue to obtain primary input for PART compliance by using expert review, under the aegis of its Board of Scientific Counselors or its Science Advisory Board. Expert review provides an independent forum for evaluation of research and complements the efforts of program managers in reviewing research activities and judging them against multi-year plans and anticipated outcomes. The expert-review panel can use intermediate outcomes to focus on key steps in the progress of any research program and to fill gaps in the spectrum of research results between outputs and ultimate outcomes. The panel's review of quality, relevance, and performance will include judgments on process efficiency and investment efficiency that should be appropriate and sufficient for the annual PART process.

The qualitative emphasis of expert review should not take away from the importance of quantitative metrics, which expert-review panels should use whenever possible to evaluate the efficiency of research processes. Examples of such processes are administration, construction, grant administration, and facility operation, in which many activities can be measured quantitatively and linked to milestones.

Process efficiency should be evaluated in the context of the expert review, but only after the relevance, quality, and effectiveness of a research program have been evaluated.

---

[7]*Performance* is described in terms of both effectiveness, meaning the ability to achieve useful results, and efficiency, meaning the ability to achieve research quality, relevance, and effectiveness with little waste.

Recommendation 3

The efficiency of research programs at EPA should be evaluated according to the same overall standards used at other agencies.

Some of the metrics proposed by EPA to comply with questions 3.4 and 4.3 of PART, such as the number of publications per full-time equivalent, have been rejected by OMB but accepted when proposed by other agencies. OMB has encouraged EPA to apply the common technique of earned value management (EVM). However, no other agency has used EVM to measure basic research.

In the committee's view, some agencies have addressed the PART questions with different approaches that are often not in alignment with their long-term strategies or missions. Many of the approaches refer only to individual portions of programs, quantify activities that are not research activities, or review processes that are not central to an agency's R&D programs. In short, many federal agencies have addressed the relevant PART questions with responses that are not, in the wording of the charge, "sufficient."

The committee calls on EPA and other agencies to address PART through consistent government-wide standards and practices addressed in its recommendations above.

ADDITIONAL RECOMMENDATION FOR THE OFFICE OF MANAGEMENT AND BUDGET

OMB should have oversight and training programs for budget examiners to ensure consistent and equitable implementation of PART in the many agencies that have substantial R&D programs.

Evaluating different agencies by different standards is undesirable because results are not comparable and ratings may not be equitable. OMB budget examiners bear primary responsibility for working with agencies in PART compliance and interpreting PART questions for them. Although the examiners cannot be expected to bring scientific expertise to their discussions with program managers, they should bring an understanding of the research process as it is performed in the context of federal agencies.

OMB decisions about whether to accept or reject metrics for evaluating the efficiency of research programs have been inconsistent. A decision to reject a given metric proposed by one agency and to accept it when proposed by another agency can unfairly damage the reputation of the first agency and diminish the credibility of the evaluation process itself. Because the framework of PART is virtually the same for all agencies and because the principles of scientific inquiry do not vary among disciplines, the implementation of PART should be both consistent and equitable in all federal research programs.

GENERAL CONCLUSIONS

The committee concluded that at the time of this study no agency had found a method of evaluating the efficiency of research based on the ultimate-outcomes of that research. Most of the methods proposed by agencies to measure efficiency addressed only particular aspects of research processes but not the research itself. In the committee's terminology, this means that agencies are focusing on process efficiency and not on investment efficiency.

The committee also concluded that sound evaluation of research should not over-emphasize efficiency, as reflected in the charge questions. The primary goal of research is knowledge, and the development of new knowledge depends on so many conditions that its efficiency must be evaluated in the context of quality, relevance, and effectiveness in addressing current priorities and anticipating future R&D questions. The criterion of relevance and timely application of the outputs from R&D in ORD and in certain program offices to the regulatory process is particularly important at an agency like EPA.

1

Introduction: The Government Performance and Results Act, the Program Assessment Rating Tool, and the Environmental Protection Agency

Federal administrations have long attempted to improve alignment of the spending decisions of the U.S. federal government with the expected results of the decisions (OMB 2004). In the 1990s, Congress and the executive branch devised a statutory and management framework to strengthen the performance and accountability of all federal agencies; the Government Performance and Results Act (GPRA) of 1993 was its centerpiece.

GPRA focused agency and oversight attention on the performance and results of government activities by requiring federal agencies to measure and report annually on the results of their activities. For the first time, each of about 1,000 federal programs was required to explicitly identify metrics and goals for judging its performance and to collect information each year to determine whether it was meeting the goals (OMB 2004). GPRA required agencies to develop a strategic plan that set goals and objectives for at least a 5-year period, an annual performance plan that translated the goals into annual targets, and an annual performance report that demonstrated whether the targets were met (NRC 1999). A key objective of GPRA was to create closer and clearer links between the process of allocating limited resources and the expected results to be achieved with them (Posner 2004).

INHERENT DIFFICULTIES IN EVALUATING RESEARCH

As agencies developed strategies to comply with GPRA, it became clear that the evaluation of science and technology research programs, especially those involving basic research, created challenges for both the agencies and oversight bodies. That, of course, is true for many other fields and practices. In the particular case of science, especially basic research, a fundamental challenge is that the course of research cannot be planned or known in advance; research entails continual feedback from observation and experimentation, which leads to new directions.

As Donald Stokes has written, "research proceeds by making choices. Although the activities by which scientific research develops new information or knowledge are exceedingly varied, they always entail a sequence of decisions or choices." They include the choice of a problem, construction of theories or models, development of instruments or metrics, and design of experiments or observations (Stokes 1997). Stokes wrote that the defining quality of basic research is that it seeks to widen the understanding of phenomena in a scientific field. In any search for new understanding, the researcher cannot know in advance what that understanding will be and therefore cannot know how long it will take, how much it will cost, and what instrumentation will be required; so the ability to evaluate progress against benchmarks is slight. Applied research is similar to basic research in that it has the same underlying process of inquiry, but it is often distinct from basic research in emphasizing the extension of fundamental understanding to "some individual or group or societal need or use" (Stokes 1997). The intended outcomes of applied research, which include methods development and monitoring, are usually well known in advance.

The committee believes that the terms *basic research*, *applied research*, and *development* describe overlapping and complementary activities. The process of research might be visualized as the development and flow of knowledge within and across categorical boundaries through collaboration, feedback loops, and fortuitous insights. Agencies support many levels of research to sustain a needed flow of knowledge, respond quickly to current demands, and prepare for future challenges.

Attempts to evaluate research in terms of efficiency may founder because of the very nature of research. Thus, a negative or unexpected result of a scientific test can have value even if the time and other resources consumed by the test might be judged "inefficient" by some metrics. In addition, much of the work of researchers involves building on, integrating, and replicating previous results and this might also appear "inefficient."

RESEARCH TERMS AT THE ENVIRONMENTAL
PROTECTION AGENCY

The Environmental Protection Agency (EPA) uses a particular nomenclature to describe its research, including the terms *core research* and *problem-driven research*. Those terms were coined by a National Research Council committee that recommended "that EPA's research program maintain a balance between problem-driven research, targeted at understanding and solving particular identified environmental problems and reducing the uncertainties associated with them, and core research, which aims to provide broader, more generic information to help improve understanding relevant to environmental problems for the present and the future" (NRC 1997). The report added that the distinction at EPA between core and problem-driven research is not always clear.[1]

EVALUATING RESEARCH UNDER THE GOVERNMENT PERFORMANCE AND RESULTS ACT

In the late 1990s, the National Academies was asked for advice on how to evaluate the research programs of federal agencies, and the Committee on Science, Engineering, and Public Policy (COSEPUP) undertook a study (NRC 1999) that began with a review of how federal research agencies were addressing GPRA. That committee determined that

• The useful outcomes of basic research cannot be measured directly on an annual basis because their timing and nature are inherently unpredictable.

• Meaningful criteria do exist by which the performance of basic research can be evaluated while the research is in progress: quality, relevance, and, when appropriate, leadership, as measured in the context of international standards.

• Such evaluations are best performed by "expert-review panels," which, in addition to experts in the field under review, include experts in related fields who may be drawn from academe, industry, government, and other appropriate sectors.

• Measurements based on those criteria can be reported regularly to assure the nation a good return on its investments in basic research.

Two years later, when more information about agencies' efforts to comply with GPRA was available, a panel appointed by COSEPUP reiterated and expanded on the original recommendations.[2] The panel focused on the five agencies that provide the majority of federal funding for research[3] and found that all had made good-faith efforts to comply with the requirements of GPRA. It also determined that some oversight bodies and agencies needed clearer procedures to validate agency evaluations and that compliance techniques, communication

---

[1]The report also pointed out that "the terms were not the same as basic vs applied research, fundamental vs directed research, or short-term vs long-term research, which are typically used by other federal agencies and researchers."

[2]In addition, COSEPUP has extended experience since the time of GPRA in helping OMB to interpret the application of government-wide criteria to agency research programs. Workshops were organized by COSEPUP for the OMB in 2001-2002 on the R&D Investment Criteria, and in 2004 on the PART.

[3]The five agencies are the Department of Defense, the Department of Energy, the National Aeronautics and Space Administration, the Department of Health and Human Services (specifically, the National Institutes of Health), and the National Science Foundation.

with oversight bodies, and timing requirements of agencies varied widely (NRC 2001).

THE RATIONALE AND FUNCTION OF THE PROGRAM ASSESSMENT RATING TOOL

Although GPRA increased the information on the results and performance of federal agencies, the Office of Management and Budget (OMB) wanted to improve the usefulness of the process and therefore developed the Program Assessment Rating Tool (PART) in 2002. PART was designed for use throughout the federal government in the larger context of "performance budgeting" and "performance measurement." Performance budgeting seeks to design budgeting procedures that optimize efficiency and effectiveness, including the most effective mechanisms for allocating available resources and holding program managers accountable for results. The first draft of PART was released for comment in May 2002 (OMB 2004), and OMB announced its intention to review each federal program every 5 years and to complete the first cycle in 2007 (OMB 2004).

PART itself is a questionnaire of at least 25 questions, whose number varies slightly with the type of program being evaluated.[4] The questions are arranged in four categories by which programs are assessed: purpose and design, strategic planning, management, and results and accountability. The answers to the questions result in a cumulative numeric score of 0-100 (100 is the best). Depending on that score, the program's rating is "effective," "moderately effective," "adequate," or "ineffective." Programs that have not developed acceptable performance metrics or sufficient performance data generally receive a rating of "Results not demonstrated." Box 1-1 shows the distribution of PART scores for the 1,016 programs rated.

PART constitutes an important step in program assessment. As one General Accountability Official told Congress, "PART may mark a new chapter in performance-based budgeting by more successfully stimulating demand for this information—that is, using the performance information generated through GPRA's planning and reporting processes to more directly feed into executive branch budgetary decisions" (Posner 2004).

The process of PART implementation is still new, and agencies are still developing their compliance methods. Some 19% of federal programs are rated "Results not demonstrated" (OMB 2007a).

THE APPLICATION OF THE PROGRAM ASSESSMENT RATING TOOL TO RESEARCH

For evaluating research programs, the PART system adopted the criteria of quality and relevance suggested by National Academies reports. Those crite-

---

[4]See Appendix C for the full list of PART questions.

ria are addressed in multiple PART questions, especially the first two sections, which explore program purpose and design and strategic planning. In lieu of the "leadership" criterion, OMB developed a new criterion of "performance."[5] This is described in terms of both effectiveness (the ability to achieve useful results) and efficiency[6] (the ability to achieve results with little waste). OMB states in its most recent PART guidance document that "because a program's performance goals represent its definition of success, the quality of the performance goals and actual performance [in achieving] those goals are the primary determinants of an overall PART rating" (OMB 2007b, p. 7). An annual retrospective analysis of results is also required.[7] Box 1-2 indicates how PART questions are scored. Reports and scores for EPA ORD programs can be found in OMB (2007a).

It appears, from the outset of its planning for PART, OMB recognized that research programs "would pose particular challenges for performance assessments and evaluations. For instance, in both applied and basic research, projects take several years to complete and require more time before their meaning for the field can be adequately understood and captured in performance reporting systems" (Posner 2004).

BOX 1-1 Distribution of PART Scores

Rating	Percent and Number
Effective	19% (186)
Moderately effective	31% (319)
Adequate	29% (289)
Ineffective	3% (27)
Results not demonstrated	19% (195)

According to OMB, "[PART] assumes that a program that cannot demonstrate positive results is no more entitled to funding, let alone an increase, than a program that is clearly failing." The consequences of a failing PART grade therefore would be nontrivial for a public agency, especially such a regulatory agency as EPA, whose actions take place in a political context.

Source: OMB 2007a.

---

[5]This distinction drew from the Army Research Laboratory's evaluation criteria of quality, relevance, and productivity.

[6]Efficiency is the ratio of the outcome or output to the input of any program (OMB 2006).

[7]According to OMB (2007b, p. 85), "programs must document performance against previously defined output and outcome measures, including progress toward objectives, decisions, and termination points or other transitions."

BOX 1-2 How PART Questions Are Scored

The four assessment categories are weighted according to the following scheme:

- Program purpose and design: 20% (5 questions).
- Strategic planning: 10% (8 questions).
- Program management: 20% (11 questions for R&D programs).
- Results and accountability: 50% (5 questions).

As a default, individual questions within a category are assigned equal weighting that total 100% for each section. However, weighting may be altered to emphasize key factors of the program.

In its 2004 budget statement, OMB wrote that the difficulty was most relevant to its preferred evaluation approach, which was to measure the outcomes[8] of research (OMB 2004): "It is preferable to have outcome measures, but such measures are often not very practical to collect or use on an annual basis. The fact is there are no 'right' measures for some programs. Developing good measures is critical for making sure the program is getting results and making an impact."

To assist agencies with significant research programs, additional instructions were added to the PART guidance and titled the "Research and Development Program Investment Criteria." The R&D Investment Criteria are found in Appendix C of the PART instructions (see Appendix G). The main body of the PART instructions applies to all federal agencies and programs, including those that perform R&D; the R&D Investment Criteria attempt to clarify OMB's expectations for R&D programs. However, a shortcoming of the Investment Criteria is that the section on Performance does not use the word efficiency, so that agencies have had to extrapolate from other sections in the guidance in evaluating that criterion.

THE ORGANIZATION AND PERFORMANCE OF
RESEARCH AND DEVELOPMENT AT THE
ENVIRONMENTAL PROTECTION AGENCY

Although PART applies to all federal research programs, the present report is concerned specifically with EPA's experience under PART. That agency has experienced difficulties in achieving PART compliance for its R&D activi-

---

[8]Outcomes may be defined as the results of research that have been integrated, assessed, and given regulatory or otherwise practical shape through a variety of actions. For example, an outcome of research on particulate matter may be improved its air quality.

ties. This section briefly reviews the purpose and organization of those activities.

EPA is primarily a regulatory agency charged with developing regulations that broadly affect human health and the environment, but its regulatory actions are intended to be based on the best possible scientific knowledge as developed both within and outside the agency. Like any other agency, EPA cannot generate all the research it needs, but several previous National Research Council reports have underscored the importance of maintaining an active and credible program of internal research (NRC 2000b, 2003). A 1992 EPA report also states that

> science is one of the soundest investments the nation can make for the future. Strong science provides the foundation for credible environmental decision making. With a better understanding of environmental risks to people and ecosystems, EPA can target the hazards that pose the greatest risks, anticipate environmental problems before they reach a critical level, and develop strategies that use the nation's, and the world's, environmental protection dollars wisely (EPA 1992).

Research at federal agencies, like other activities, is organized to support an agency mission. EPA's process is described in its strategic plan. EPA drew up its first strategic plan in 1996 in response to GPRA. That plan, which has been renewed every 3 years, stated that "the mission of the U.S. Environmental Protection Agency is to protect human health and to safeguard the natural environment—air, water, and land—upon which life depends" (EPA 1997a). The current strategic plan (2006-2011) [EPA 2006] has five principal goals, all of which have scientific underpinnings:

- Clean air and addressing global climate change.
- Clean and safe water.
- Land preservation and restoration.
- Healthy communities and ecosystems.
- Compliance and environmental stewardship.

The plan also lists three "cross-goal strategies" that describe values meant to guide planning for all five goals: results and accountability, innovation and collaboration, and best available science (see Appendix D).[9]

Research-related activities at EPA, both internal and external, are the responsibility of the Office of Research and Development (ORD) as well as EPA's program offices and regional laboratories. The committee chose to focus its review on ORD's research program as this is where the controversy regarding the

---

[9]See Appendix D for more information about EPA's strategic planning and multi-year planning process.

development of its efficiency measures arose.[10] The ORD conducts research in its in-house laboratories, develops risk-assessment methods and regulatory criteria, and provides technical services in support of the agency's mission and its program and regional offices. It is organized in three national laboratories, four national centers, and two offices in 14 facilities around the country and its headquarters in Washington, DC. ORD also has an extramural budget for grants, cooperative and interagency agreements, contracts, and fellowships that has accounted for 40-50% of its total budget in recent years.[11] Its first strategic plan set forth a straightforward vision (EPA 1996, 1997b): "ORD will provide the scientific foundation to support EPA's mission."

In 1999, ORD began to organize its scientific activities in multi-year plans to improve continuity and strategic integration. The multi-year plans, typically covering 5 years, are developed by research teams in ORD laboratories and centers and are peer-reviewed. They cover 16 subjects (such as drinking water, safe food, and ecologic research) and are updated annually (EPA 2008).

USES OF RESULTS OF THE ENVIRONMENTAL
PROTECTION AGENCY RESEARCH

The results of EPA research are used both by the agency itself and by various others outside the agency. The explicit purpose of both ORD research and extramural research is to provide scientific bases of EPA actions. The research may lead to end outcomes when results are integrated, assessed, and given regulatory or otherwise practical shape through actions in or outside EPA.

Another important goal of EPA's work, however, is to provide knowledge outputs for diverse organizations that have environmental interests and responsibilities, such as state and local governments, nongovernment organizations, international organizations, and community groups. Such entities interpret and use ORD outputs for their own planning and regulatory purposes.

SUMMARY

In the last decade and a half, two important efforts to evaluate the work of government agencies have been developed: GPRA, passed into law by Congress in 1993, and PART, designed in 2002 by OMB and first applied in 2003. Both efforts apply to all federal programs, including R&D programs. Evaluating the efficiency of R&D, required under PART, has proved to be challenging for all

---

[10]For this reason, the committee did not address economic analysis, as this work is not conducted by ORD. A report summarizing the history of economics research at EPA through the 1990's can be found at: http://yosemite.epa.gov/ee/epa/eed.nsf/webpages/EconomicsResearchAtEPA.html.

[11]For the FY 2007 enacted budget, 43% ($239,168,600 of $555,383,000) was budgeted to be spent extramurally. [EPA, ORD, personal communication, 2007]

research-intensive agencies.[12] One of those is EPA, which has sought the assistance of the National Academies in its effort to comply with the efficiency questions of PART. The next chapter examines in some detail the complex process of PART compliance and how various agencies have responded.

REFERENCES

EPA (U.S. Environmental Protection Agency). 1992. Safeguarding the Future: Credible Science, Credible Decisions. The Report of the Expert Panel on the Role of Science at EPA. EPA/600/9-91/050. U.S. Environmental Protection Agency, Washington, DC.

EPA (U.S. Environmental Protection Agency). 1996. Strategic Plan for the Office of Research and Development. EPA/600/R-96/059. Office of Research and Development, U.S. Environmental Protection Agency, Washington, DC.

EPA (U.S. Environmental Protection Agency). 1997a. EPA Strategic Plan. EPA/190-R-97-002. Office of the Chief Financial Officer, U.S. Environmental Protection Agency, Washington, DC.

EPA (U.S. Environmental Protection Agency). 1997b. Update to ORD's Strategic Plan. EPA/600/R-97/015. Office of Research and Development, U.S. Environmental Protection Agency, Washington, DC.

EPA (U.S. Environmental Protection Agency). 2006. 2006-2011 EPA Strategic Plan: Charting Our Course. Office of the Chief Financial Officer, U.S. Environmental Protection Agency. September 2006 [online]. Available: http://www.epa.gov/cfo/plan/2006/entire_report.pdf [accessed Feb. 7, 2008].

EPA (U.S. Environmental Protection Agency). 2008. Research Directions: Multi-Year Plans. Office of Research and Development, U.S. Environmental Protection Agency [online]. Available: http://www.epa.gov/ord/htm/aboutord.htm [accessed Feb. 7, 2008].

NRC (National Research Council). 1997. Building a Foundation for Sound Environmental Decisions. Washington, DC: National Academy Press.

NRC (National Research Council). 1999. Evaluating Federal Research Programs: Research and the Government Performance and Results Act. Washington, DC: The National Academies Press.

NRC (National Research Council). 2000a. Experiments in International Benchmarking of U.S. Research Fields. Washington, DC: National Academy Press.

NRC (National Research Council). 2000b. Strengthening Science at the U.S. Environmental Protection Agency. Washington, DC: National Academy Press.

NRC (National Research Council). 2001. Pp. 2-3 in Implementing the Government Performance and Results Act for Research Programs: A Status Report. Washington, DC: The National Academies Press.

NRC (National Research Council). 2003. The Measure of STAR: Review of the U.S. Environmental Protection Agency's Science to Achieve Results (STAR) Research Grants Program. Washington, DC: The National Academies Press.

---

[12]In this report, *research-intensive* is used to describe agencies of whose mission's research is an essential, though not necessarily dominant, aspect. For example, research is important at EPA but not the primary function of the agency, as it is for the National Institutes of Health or the National Science Foundation.

OMB (Office of Management and Budget). 2004. Pp. 49-52 in Rating the Performance of Federal Programs. The Budget for Fiscal Year 2004. Office of Management and Budget [online]. Available: http://www.gpoaccess.gov/usbudget/fy04/pdf/budget/performance.pdf [accessed Nov. 7, 2007].

OMB (Office of Management and Budget). 2007a. ExpectMore.gov. Office of Management and Budget [online]. Available: http://www.whitehouse.gov/omb/expect more/ [accessed Nov. 7, 2007].

OMB (Office of Management and Budget). 2007b. Guide to the Program Assessment Rating Tool (PART). Office of Management and Budget. January 2007 [online]. Available: http://stinet.dtic.mil/cgi-bin/GetTRDoc?AD=ADA471562&Location= U2&doc=GetTRDoc.pdf [accessed Nov. 7, 2007].

Posner, P.L. 2004. Performance Budgeting: OMB's Performance Rating Tool Presents Opportunities and Challenges for Evaluating Program Performance: Testimony before the Subcommittee on Environment, Technology, and Standards, Committee on Science, House of Representatives, March 11, 2004. GAO-04-550T. Washington, DC: U.S. General Accounting Office [online]. Available: http://www.gao.gov/new.items/d04550t.pdf [accessed Nov. 7, 2007].

Stokes, D.E. 1997. Pp. 6-8 in Pasteur's Quadrant: Basic Science and Technological Innovation. Washington, DC: Brookings Institution.

2

Efficiency Metrics Used by the Environmental Protection Agency and Other Federal Research and Development Programs

The questions in the Program Assessment Rating Tool (PART) address many aspects of programs of the federal government, but the charge to this committee refers specifically to questions used to evaluate efficiency. The charge asks

1. What efficiency measures are currently used for EPA R&D programs and other federally-funded R&D programs?
2. Are these efficiency measures sufficient? Outcome based?
3. What principles should guide the development of efficiency measures for federally-funded R&D programs?
4. What efficiency measures should be used for EPA's basic and applied R&D programs?

This chapter addresses primarily the first question in the charge. To answer that question, the committee examined many of the efficiency metrics proposed to comply with PART by EPA and other federal agencies engaged in research. The committee reviewed documents, interviewed agency personnel, and heard presentations during a workshop in April 2007 that was attended by most of the research-intensive agencies and several large corporations that emphasize research.[1] This chapter summarizes some of the challenges of evaluating research efficiency and ways in which agencies have approached those challenges.

---

[1]The workshop was held on April 24, 2007, at the National Academies, 2101 Constitution Avenue, Washington, DC 20418. A workshop summary appears in Appendix B.

EVALUATING RESEARCH AND DEVELOPMENT

Research is difficult to evaluate by any mechanism. Useful evaluation requires substantial elapsed time because research on a given scientific question may span 3-5 years from initiation of laboratory or field experiments to analysis and publication of results. Substantial time may also be required for training of EPA staff or the scientific community in scientific and technical advancements prior to the conduct of the research. Considerably more time may elapse before the broader impacts of published research are apparent (NRC 2003).

Although the committee was asked specifically to render advice on how EPA could best comply with the efficiency questions of PART, it concluded that more general suggestions on the evaluation of research would also have value for research-intensive agencies, for the Office of Management and Budget (OMB) and the Office of Science and Technology Policy (OSTP), and for Congress. It therefore examined the particular details of PART (Chapters 2-4), proposed principles that can be used to evaluate the results of research in any federal agency, and provided recommendations for EPA that other agencies also may find useful (Chapter 5).

THE PROGRAM ASSESSMENT RATING TOOL AND EFFICIENCY

Efficiency is a common enough concept, as illustrated by familiar dictionary definitions: "effective operation as measured by a comparison of production with cost (as in energy, time, and money)" and "the ratio of the useful energy delivered by a dynamic system to the energy supplied to it."[2] The PART approach to efficiency is explained this way by OMB (OMB 2007a, p. 9):

> Efficiency measures reflect the economical and effective acquisition, utilization, and management of resources to achieve program outcomes or produce program outputs. Efficiency measures may also reflect ingenuity in the improved design, creation, and delivery of goods and services to the public, customers, or beneficiaries by capturing the effect of intended changes made to outputs aimed to reduce costs and/or improve productivity, such as the improved targeting of beneficiaries, redesign of goods or services for simplified customer processing, manufacturability, or delivery.

APPLYING EFFICIENCY TO INPUTS, OUTPUTS, AND OUTCOMES

Any definition of efficiency depends on the process to which it is applied. Of relevance to this report is its application to the processes of research and development, which, as described by OMB, are complex and variable and involve

---

[2]Merriam Webster Online, http://www.m-w.com/.

inputs, outputs, and outcomes that vary by agency, program, and laboratory. Those terms are discussed by OMB (OMB 2007a) as follows:

- *Inputs* for purposes of PART are any agency resources that support research, which may include "overhead, intramural/extramural spending, infrastructure, and human capital" (OMB 2007a, p. 76).
- *Outputs* "describe the level of activity that will be provided over a period of time, including a description of the characteristics (e.g., timeliness) established as standards for the activity. Outputs refer to the internal activities of a program (i.e., the products and services delivered)" (OMB 2007a, p. 83). Outputs that have been used by agencies to comply with PART include research findings, papers published or cited, grants awarded, adherence to a projected schedule, and variance from cost and time schedules (OMB 2007b).
- *Outcomes,* according to OMB guidance, "describe the intended result of carrying out a program or activity. They define an event or condition that is external to the program or activity and that is of direct importance to the intended beneficiaries and/or the public" (OMB 2007a, p. 8). OMB gives the example of a tornado-warning system, whose outcomes "could be the number of lives saved and property damage averted" (OMB 2007a, p. 9). An outcome of research in support of the mission of a regulatory agency, such as EPA, may be the consequence of a regulation or other change that brings about some improvement in health or environmental quality.

THE PROGRAM ASSESSMENT RATING TOOL GRADING SYSTEM

The overall structure of PART was introduced in Chapter 1; additional aspects are described briefly here.

Although this report focuses on the criterion of efficiency, the PART questions, taken in their entirety, attempt to address a broad range of issues with the goal of a thorough evaluation of federal programs. For example, section 1 of PART asks for general information about purpose and design. The second section, on strategic planning, asks such questions as whether a program has a long-term plan (2.1) and annual plans (2.2), both of which concern the relevance of research to the agency mission; it then asks whether the program is meeting long-term targets (2.3) and annual targets (2.4) concerning the quality of the research. The requirements for annual plans and annual targets often present problems for research managers. As discussed in Chapter 1, programs engaged in core research may be unable to specify the nature, timing, or benefits of their work annually because the results of core research can seldom be planned or analyzed in 1-year increments.

Many of these sections also assume that evaluation be based on ultimate outcomes. For example, the PART guidance for question 2.2 *(Does the program have ambitious targets and timeframes for its long-term measures?)* states as follows: "For R&D programs, a *Yes* answer would require that the program provides multi-year R&D objectives. Where applicable, programs must provide schedules with annual milestones, highlighting any changes from previous schedules. Program proposals must define what outcomes would represent a minimally effective program and a successful program." [Additional examples of the emphasis on ultimate outcomes can be seen in the sections of the guidance included in Appendix I.]

In Section 3, which deals with management, question 3.4 specifies quantitative efficiency metrics as follows: "Does the program have procedures (e.g., competitive sourcing/cost comparison, IT improvements, and appropriate incentives) to measure and achieve efficiencies and cost effectiveness in program execution?" To win approval, a program must have "regular procedures to achieve efficiencies and cost effectiveness and at least one efficiency measure that uses a baseline and targets" (OMB 2007a, p. 26).

Question 4.3 also addresses efficiency but appears in Section 4, on results and accountability: "Does the program demonstrate improved efficiencies or cost effectiveness in achieving program goals each year?" To pass question 4.3, a program must have passed question 3.4 (that is, have "at least one efficiency measure that uses a baseline and targets") and have demonstrated improved efficiency or cost effectiveness over the prior year.

In discussions with the committee, OMB officials were consistent in supporting the use of outcome-based measures to evaluate efficiency. They also acknowledged that efforts to do so had not yet been successful.[3]

Because agencies are expected to provide a satisfactory answer to every PART question and several questions require answers on efficiency and annual achievements, this focus on efficiency and annual achievements can lead to a poor PART grade even if the relevance, quality, and effectiveness of a research program are demonstrated. That and other difficulties are addressed further in Chapter 3.

THE USE OF "EXPERT REVIEW" AT THE
ENVIRONMENTAL PROTECTION AGENCY

EPA uses multiple mechanisms to evaluate its R&D activities, including internal processes of strategic plans, multi-year plans, and annual performance goals (mentioned in Chapter 1). The multi-year plans provide a means for tracking and, when necessary, adjusting research activities as they progress toward long-term goals.

---

[3]As one example, the committee heard the following from OMB: "The requirement for 3.4 is that they have an efficiency measure. The highest standard is outcome. However, if that is not achievable, then they can have an output efficiency measure... But we are pushing for the outcomes, because if we just focus on the activities that we do, we don't necessarily have the ability to find out if those activities and strategies are effective. That's why another key component of the PART is evaluation" (from April 2007 workshop discussion; see summary in Appendix B).

To gain independent external perspective, EPA uses several standing "expert review" boards. Expert review is a broadened version of peer review, the mechanism by which researchers' work is traditionally judged by other researchers in the same field.[4] Expert review groups may include not only experts in the field under review but members from other fields and appropriate "users" of research results, who may represent the private sector, other agencies, non-government organizations (NGOs), state governments, labor unions, and other relevant bodies (NRC 1999).

At EPA, expert reviewers are chosen for their skills, experience, and ability to judge not only the quality, relevance, and effectiveness of a program but whether it is being efficiently planned, managed, and revised in response to new knowledge—that is, whether it is efficient. Each panel should include members who have successfully run research programs themselves and are able to recognize good performance.

The Science Advisory Board

One of EPA's long-standing review bodies is the Science Advisory Board (SAB), which includes a mix of scientists and engineers in academe, industry, state government, advisory bodies, and NGOs. The SAB was established by Congress in 1978 under a broad mandate to advise the agency on technical matters, including the quality and relevance of information used as the basis of regulations. The panel includes experts in science and technology policy, environmental-business planning processes, environmental economics, toxicology, resource management, environmental decision-making, ecotoxicology, risk perception and communication, decision analysis, risk assessment, civil and environmental engineering, epidemiology, radiologic health, air-quality modeling, public health, and environmental and occupational health (EPASAB 2007).

According to the *Overview of the Panel Formation Process at the Environmental Protection Agency Science Advisory Board* (EPASAB 2002), EPA uses the following criteria in evaluating an individual panelist to serve on the SAB:

- Expertise, knowledge, and experience (primary factors).
- Availability and willingness to serve.

---

[4]According to one definition, "peer review is a widely used, time-honored practice in the scientific and engineering community for judging and potentially improving a scientific or technical plan, proposal, activity, program, or work product through documented critical evaluation by individuals or groups with relevant expertise who had no involvement in developing the object under review" (NRC 2000). Expert review was also recommended by the Committee on Science, Engineering, and Public Policy panel cited in Chapter 1 for evaluating research to comply with the Government Performance and Results Act.

- Scientific credibility and impartiality.
- Skills working in committees and advisory panels.

The Board of Scientific Counselors

The other principal EPA expert-review panel for ORD is the Board of Scientific Counselors (BOSC), a body of nongovernment scientists and engineers established in 1996 to provide advice, information, and recommendations to ORD. It has up to 15 members, and they meet three to five times a year. The BOSC reviews are relevant to this discussion because EPA is experimenting with their use as a mechanism for reviewing various aspects of research effectiveness.

In 2004, the BOSC review process was restructured to focus on the three evaluation criteria of PART and to include both prospective and retrospective reviews of research programs. In 2006, three charge questions were developed for use in BOSC's summary assessment of each program's long-term goals:

- How appropriate is the research used to achieve each long-term goal? Is the program still asking the right questions, or have they been superseded by advancements in the field? *(Relevance)*
- How good is the technical quality of the program's research products? *(Quality)*
- How much are the program results being used by environmental decision-makers to inform decisions and achieve results? *(Performance)*

The BOSC review process also feeds into PART through several other questions. For example, the BOSC review is submitted in response to question 4.5, "Do independent evaluations of sufficient scope and quality indicate that the program is effective and achieving results?"[5]

In spring 2006, a BOSC panel added the charge questions discussed above to its evaluation criteria; it used them for the first time early in 2007. Although there is much overlap between the BOSC investigations and the PART questions regarding results of a program, OMB had not by the time of this committee's investigation determined whether the use of BOSC's revised charge met the requirements of PART.[6]

During the 2005 PART review for EPA's drinking-water program, EPA experimented with a quantitative version of the BOSC evaluation. It involved nine questions, for each of which the committee gave a rating of 1-5 to provide a

---

[5]For additional material about BOSC, see EPA's *Draft Board of Scientific Counselors Handbook for Subcommittee Chairs*, Appendix B, p. 18.

[6]During the July 2007 workshop, committee members discussed this issue with representatives of OMB and EPA. The EPA representatives described current efforts to develop a quantitative system for use by BOSC.

numerical grade. That process was not accepted as scientifically valid by the BOSC.

Two BOSC reviews were in progress at the time of this report: on pesticides and toxics and on sustainability research. EPA staff noted that the reviews will serve as baselines for later reviews. EPA is providing the BOSC review committees with two kinds of data, among others, to evaluate the performance of research programs: pilot surveys that evaluate how the research is being used and bibliometric analyses (P. Juengst, EPA, personal communication, 2007). Box 2-1 provides an example of a BOSC expert review. According to EPA staff, there is increasing pressure from OMB to focus on outcome-based efficiency metrics, but the agency has been unable to establish such metrics for research.

EMERGING ISSUES

One important role of expert review is to complement the ability of program managers and agency leaders to anticipate important emerging issues. Strategic effectiveness rises when the agency plans for the "next big thing," rather than awaiting its sudden arrival. The program managers necessarily focus their attention on the day-to-day demands of administration, but expert reviewers can survey agency research in a wider context. To the degree that an agency can position itself at the forefront of a new field, it can increase its research relevance, quality, and performance.

METRICS PROPOSED BY THE
ENVIRONMENTAL PROTECTION AGENCY

EPA, like other agencies, proposed quantitative metrics to measure various kinds of efficiency in its PART compliance, and many of them were accepted by OMB. Metrics considered or used by EPA included the use of its research results to support regulations, surveys to gauge client satisfaction with its products, average time spent in producing assessments, overhead as a fraction of research, and citations per dollar invested. Such metrics fit well with the existing strategic and multi-year planning that provides annual milestones against which to evaluate them.

Like other agencies, EPA has proposed that an increase in the number of peer-reviewed publications produced per full-time equivalent (FTE) complies with the PART guidance that "efficiency measures could focus on how to produce a given output level with fewer resources" (OMB 2006a, p. 10). That was not accepted by OMB examiners for question 3.4 as an efficiency metric for the Water Quality Research Program. OMB found that the lack of a tight linkage between publications and budget made it hard to determine whether money was being spent appropriately. Publications might have been far ahead of schedule and over budget, for example, or behind schedule and under budget (K. Neyland, OMB, personal communication, 2007).

BOX 2-1 BOSC: An Example of Expert Review

One example of the composition and function of a BOSC expert-review panel is its Human Health Subcommittee, which issued a report on EPA's Human Health Research Program (HHRP) in 2005. The panel included eight members in academe, industry, and government.[7] The panel met for 3 days and stated its purpose as follows: "The objective of this review is to evaluate the relevance, quality, performance, and scientific leadership of the Office of Research and Development's (ORD's) Human Health Research Program."

It evaluated the overall program's relevance, quality, performance, and leadership relative to each of its four long-term goals:

- Use of mechanistic data in risk assessment.
- Aggregate and cumulative risk assessment.
- Evaluation of risk to susceptible subpopulations.
- Evaluation of public-health outcomes.

The subcommittee visited the HHRP's main facility, in Research Triangle Park, North Carolina, where it heard from EPA offices and programs regarding the utility of research products developed by ORD scientists in the HHRP.

The expert-review panel received extensive confirmation that ORD scientists were helpful to the various EPA regions in hosting regional scientists in ORD laboratories, collaborating with the regions on regional environmental problems, providing scientific consultation to the regions to help to ameliorate their environmental problems, and providing scientific consultation to the regions on specific problems in environmental toxicology.

Earned-Value Management

EPA staff also approached OMB to discuss the use of EVM as an efficiency metric for its ecologic research program. EVM measures the degree to which research outputs conform to scheduled costs along a timeline. It is used by agencies and other organizations in many management settings, such as construction projects and facilities operations, where the outcome (such as a new

---

[7]The members of the Human Health Subcommittee, according to the BOSC Web site, had "considerable expertise in the area of human health research, including formal education, training, and research experience in biology, chemistry, biochemistry, environmental carcinogenesis, pharmacology, molecular biology and molecular mechanisms of carcinogenicity and toxicity, toxicology, physiologically based pharmacokinetic (PBPK) modeling, exposure modeling, risk assessment, epidemiology, biomarkers and biological monitoring, and public health, with additional expertise in the areas of children's health, community-based human exposure studies, and clinical experience" (EPA 2005, p. 1).

laboratory or optimal use of facilities) is well known in advance, and progress can be plotted against milestones. Although EPA and other agencies have found value in using EVM to measure the efficiency of some processes, they have not found a way to apply it to research outcomes.

METRICS THAT DID NOT PASS THE PROGRAM ASSESSMENT RATING TOOL PROCESS

Since 2004, when PART grading began, several major research-based EPA programs have been rated "ineffective" or "results not demonstrated." One was the Ecological Research Program (ERP), which in 2005 was given an "ineffective" rating,[8] including a "no" on question 3.4. At that time, it was noted that the program lacked an acceptable efficiency measure, but was working to develop one.

In EPA's view, the agency failed to provide an acceptable efficiency metric because it could not measure the outcome efficiency of its research (M. Peacock, EPA, personal communication, 2007). It was also given a zero score on question 4.3 because a program that fails to have an acceptable efficiency metric on question 3.4 cannot demonstrate annual increases in efficiency. Other programs that have not passed the efficiency questions include the National Ambient Air Quality Standards Program and the Ground Water and Drinking Water Program, for similar reasons.

In examining OMB documents and the ExpectMore.gov Web site, the committee found that for most research-intensive agencies other than EPA, OMB accepted efficiency measures similar to those proposed by EPA, such as scheduled regulatory decision-making activities. No agency has responded to PART questions 3.4 and 4.3 by using outcome-based efficiency measures for R&D programs.

The Appeals Process

EPA appealed the denial of using publications per FTE as an efficiency metric for the Water Quality Research Program to an OMB Appeals Board, which accepted the appeal on several conditions. One was that "the program should include a follow-up action in its PART improvement plan relating to developing an outcome-oriented efficiency metric," and another was that the program "must have a baseline and targets" (OMB 2006b). Even though the use of "outcome-oriented efficiency metrics" is not required by the PART guidance or the R&D Investment Criteria, it is strongly preferred by OMB and sometimes required by the examiner and/or during the appeals process. In both EPA and other research-intensive agencies, concerns have been voiced by those involved

---

[8]PART rated 3% of federal programs "ineffective." It rated 19% as "results not demonstrated" (OMB 2007c).

in PART compliance that the application of rules sometimes seems inconsistent or confusing.[9] In addition, OMB continues to encourage development of a version of EVM that will prove satisfactory for the purpose of evaluating the efficiency of R&D programs, but at the time of the present committee's study the issue had not been resolved (B. Kleinman, OMB, personal communication, 2007).

THE CONSEQUENCES OF A "NO" ANSWER TO A PROGRAM ASSESSMENT RATING TOOL QUESTION

Because PART was initiated in 2003 and has not yet examined all research programs of federal agencies, there is little information about the effects of low ratings on agencies. However, as demonstrated above by the example of the ERP, a program may do poorly in the PART process if it does not have an acceptable measure of efficiency even if it has high marks for relevance and quality. A rating of "ineffective" for research cannot be helpful for a regulatory agency like EPA, whose authority rests in part on its reputation for sound scientific research. Indeed, after the "ineffective" PART ratings were applied to the ERP in 2005, the program suffered substantial erosion of support (Morgan 2007).[10] According to figures cited by EPA's Risk Policy Report (Sarvana 2007), Congress reduced extramural funding for "ecology and global change" through EPA's National Center for Environmental Research from about $32 million in FY 2002 to less than $25 million in FY 2003. The program received a similar cut from FY 2004, when it received about $24 million, to FY 2005, when it received only about $8 million.[11]

---

[9]According to one study of the early application of PART, "The patterns of rating programs are not very clear regarding the FY 2004 process, largely because of variability among the OMB budget examiners. The variability was pointed out by GAO in its assessment of the process" (Radin 2006, p. 123). See also comments by NASA representative on p. 33 below.

[10]In his testimony before the Subcommittee on Energy and Environment, Committee on Science and Technology, U.S. House of Representatives, March 2007, M. Granger Morgan, of Carnegie Mellon University (CMU), noted the agency's difficulty in carrying out the work necessary to comply with PART while its budget was reduced. He stated that "it appears seriously misguided to raise the bar for comprehensive cost-effective or benefit-cost justification for environmental science research, while simultaneously shrinking the resources devoted to the types of research needed to assess the net social benefits of the outcomes of environmental science research." Morgan is chair of CMU's Department of Engineering and Public Policy, chair of the EPA SAB, and an expert in risk analysis and uncertainty (Morgan 2007, p. 4).

[11]Most recently, the ERP received a positive rating, again as reported by InsideEPA.com: "The rating was part of OMB's Program Assessment Rating Tool (PART) process, which rates federal programs' performance and helps set budget levels. It is the third PART review of ERP since OMB launched the initiative in 2002, and the first to give the program a positive score. Previous PART reviews criticized ERP for not fully

An "ineffective" PART rating also affects program ratings under the President's Management Agenda (PMA). The PMA is relevant to this document, even though the committee's charge did not specify it, because PART is a component of the PMA. The PMA is generally an agency-level effort with five initiatives, one of which is Budget-Performance Integration (BPI). All programs assessed by PART must have acceptable efficiency measures for the agency to receive a "green" score in the BPI.[12]

EVALUATION MECHANISMS USED BY OTHER AGENCIES

The committee and staff have consulted with other agencies about their PART evaluation processes. They also invited research-intensive agencies' representatives to a workshop at the Academies in April 2007 to describe their processes and compare results (see Appendix B for the workshop summary). This section is based on the workshop presentations and followup conversations.

Metrics of Efficiency Accepted by the
Office of Management and Budget

OMB makes clear its general preference for "outcome efficiency" in PART compliance (OMB 2006a), but the mechanisms proposed by agencies are metrics of output (process) efficiency. The committee compiled and reviewed an "efficiency measures table" of efficiency metrics used for research programs by 11 federal agencies, including EPA, in following the PART guidance[13] (see Appendix E for Table E-1). The table also includes information gathered from four corporations that have R&D programs. The following list is a sample of the common types of metrics proposed by the agencies, many of which have been accepted by OMB:

- Time to process and award grants.
- Time to respond to information requests.
- Publications per FTE (or per dollar).
- Percentage of budget that is overhead.
- Percentage of work that is peer-reviewed.
- Average cost per measurement or analysis.
- Cost-sharing.

---

demonstrating the results of programmatic and research efforts—and resulted in ERP funding cuts" (Inside EPA Risk Policy Report 2007).

[12]The PMA awards agencies a green, yellow, or red rating. As noted in PART guidance (OMB 2007a, p. 9): "The President's Management Agenda (PMA) Budget and Performance Integration (BPI) Initiative requires agencies to develop efficiency measures to achieve *Green* status."

[13]The complete table is in Appendix E.

- Quality or cost of equipment and other inputs.
- Variance from schedule and cost.

Chapter 3 asks whether those metrics are "sufficient" for evaluating research efficiency.

Department of Energy

The Department of Energy (DOE) uses PART to evaluate many management processes of its Office of Science. For evaluating the quality and relevance of research, DOE depends on peer review of all portfolios every second or third year by "committees of visitors." That was found to be fairly cost-effective, allowing the agency to look at what was proposed and how well it was performed, to identify ideas that lack merit, to discontinue inefficient processes, to redirect R&D, or to terminate a poorly performing project.

With the creation of PART, a committee was established to test appropriate metrics, but the committee has not found a way to assign value to a basic-research portfolio. A DOE representative commented that the work is valued according to its societal and mission accomplishments; this has to be done by working closely with the scientific community.

A DOE representative said that the director of the President's Office of Science and Technology Policy had established a committee charged with developing a mechanism for measuring the value of research and estimating the cost of compliance.

National Science Foundation

Like DOE, the National Science Foundation (NSF) uses external committees of visitors to perform peer review (called merit review) on programs or portfolios every 3 years. Merit review is a detailed and long examination of technical merit and broader impacts of research.

NSF tracks efficiency primarily in two ways. One measures the time to decision on research awards; the second measures facility costs, schedules, and operations, with specific goals for each.

National Aeronautics and Space Administration

The National Aeronautics and Space Administration (NASA) uses PART exercises in evaluating the efficiency of repetitive, stable, and baseline processes and some aspects of R&D, such as financial management, contracting, travel processing, and capital-assets tracking. The agency has been using PART metrics to track and evaluate the complex launch process and to find safe ways to reduce the size of the Space Shuttle workforce. Other uses were planned, such as increasing the on-time availability and operation of ground test facilities and reducing the cost per minute of operating space network support for missions.

Like other agencies, NASA does not find PART useful for evaluating the efficiency of research, especially unrepeatable projects, such as discoveries dictated by science or the development of prototypes.

A NASA representative notes that the PART process depended heavily on the PART examiners, who tended to vary widely in attitudes and experience. Although the NASA examiners typically had scientific or engineering backgrounds, this was not necessarily the case for OMB policy-makers crafting the PART policies and guidance. The representative notes that because the OMB policy-makers generally did not have research backgrounds, NASA spends considerable time in educating them about the relevant differences in R&D programs, as these differences are not well considered in PART guidance. The representative suggests more flexibility in the actions of the reviewers—for example, in recognizing that short-term decreases in efficiency might lead to long-term efficiency gains and in seeing the need to balance efficiency with effectiveness.[14]

National Institutes of Health

National Institutes of Health (NIH) staff described using PART on research and research-support activities. For example, the extramural research program has achieved cost savings through improved grant administration. The intramural research program has used it to reallocate laboratory resources, and the building and facilities program has monitored its property condition index. The extramural construction program has achieved economies by expanding the use of electronic management tools to monitor construction and occupancy for 20-years post-completion.

NIH staff notes that the PART approach to efficiency is the same as a business model which emphasizes time, cost, and deliverables. With such a model, efficiency can be increased by improving any variable, so long as the other two do not worsen. This approach does not fit the scientific discovery process. Some 99% of the NIH portfolio has been subjected to PART; 95% percent of the programs are rated as effective, and the other 5% as moderately effective. External research (close to 90% of the budget), which is not under NIH's direct control, is excluded, although the agency coordinates with awardees to ensure performance.

---

[14]In a case study of the PART process at the Department of Health and Human Services, a similar difficulty was described. "In some cases, OMB budget examiners were willing to deal with multiple elements of programs as a package; in other cases, the examiner insisted that a small program would require individual PART submissions. It was not always clear to HHS staff why a particular program received the rating it was given; OMB policy officials did not appear to have a consistent view of the PART process" (Radin 2006, p. 142).

NIH staff notes that, in scientific discovery, variables are largely unknown because the outcome is unpredictable knowledge, and the inputs of time, cost, and resources are difficult to estimate. Research does not fit easily into a business model, for other reasons cited:

- In research, high-risk projects are strongly associated with innovative outcomes that may initially fail even though the scientific approach was sound.
- A research outcome may be unexpected or lead to an unexpected benefit.
- Changing direction, which may look like poor management in the context of a business model, may be good research practice in order to conduct good science.
- A business model does not capture the null hypothesis, or a serendipitous finding that gives valuable information (Duran 2007).

These types of research results provide valuable information, which is likely not credited in a business model approach.

National Institute for Occupational Safety and Health

The National Institute for Occupational Safety and Health (NIOSH) functions as the research partner of the Occupational Safety and Health Administration. NIOSH uses independent expert review to evaluate its 30 research programs, which exist in a "matrix" with substantial overlaps.

A model for evaluating a research program was provided by a framework developed for NIOSH by the National Academies with a blend of quantitative and qualitative elements. A central feature of the framework is that it adds to outputs and outcomes a third metric, "intermediate outcomes." The categories were associated with the following kinds of results (Sinclair 2007):

- *Outputs.* Peer-reviewed publications, NIOSH publications, communications to regulatory agencies or Congress, research methods, control technologies and patents, and training and information products.
- *Intermediate outcomes.* Regulations, guidance, standards, training and education programs, and pilot technologies.
- *End outcomes.* Reductions in fatalities, injuries, illnesses, and exposures to hazards.

The Academies' review of NIOSH procedures included several approaches to program evaluation, including the following general road map for characterizing the evaluation process as a whole (NRC 2007):

- Gather appropriate information from NIOSH and other sources.
- Determine timeframe for the evaluation.

- Identify program-subject challenges and objectives.
- Identify subprograms and major projects in the research program.
- Evaluate the program and subprogram components sequentially (this will involve qualitatively assessing each phase of a research program).
- Evaluate the research program's potential outcomes not yet appreciated.
- Evaluate and score the program's potential outcomes and important subprogram outcomes specifically for contributions to the environment and health.
- Evaluate and score the overall program for quality, using a numerical scale.
- Evaluate and score the overall program for relevance, using a numerical scale.
- Evaluate and score the overall program for performance (effectiveness and efficiency), using a numeric scale.
- Identify important emerging research.
- Prepare report.

One attribute of that approach, said a NIOSH representative, is that it is flexible enough to apply to R&D programs in which efficiency metrics are appropriate for some functions but not for others. It allows evaluation of a research program by focusing on outputs agreed to in the multi-year plan, and the outcomes are embedded in the plan itself. The plan can thus be evaluated by year-over-year results that lay the foundation for the outcomes.

The NIOSH plan could also make use of a numeric scale for performance. For example, expert-review panels could be asked to assess performance on a scale of 1-5.

METHODS USED BY INDUSTRY

Representatives of four industries that perform research participated in the committee's public workshop and described how research delivers value to their companies. They all had clear rationales for investing in research and spent substantial amounts (typically about 1% of sales) on research activities, but none used EVM, none evaluated research in terms of efficiency, and none described direct links to outcomes (sales).[15] Chapter 3 offers additional perspective on evaluating research efficiency in industry.

SUMMARY

OMB has required that federal agencies measure the efficiency of their

---

[15]See workshop summary in Appendix B.

activities to comply with PART. This chapter has summarized how agencies with substantial research programs, including EPA, have attempted to comply with that requirement. In the next chapter, we discuss whether the metrics that agencies have proposed and are using are "sufficient" for evaluating the performance of research, including its efficiency, and discuss a balanced approach to such evaluation.

REFERENCES

Duran, D. 2007. Presentation at the Workshop on Evaluating the Efficiency of Research and Development Programs at the Environmental Protection Agency, April 24, 2007, Washington, DC.

EPA (U.S. Environmental Protection Agency). 2005. Program Review Report of the Board Scientific Counselors: Human Health Research Program. Office of Research and Development, U.S. Environmental Protection Agency [online]. Available: http://www.epa.gov/osp/bosc/pdf/hh0507rpt.pdf [accessed Nov. 8, 2007].

EPASAB (U.S. Environmental Protection Agency Science Advisory Board). 2002. Overview of the Panel Formation Process at the Environmental Protection Agency. EPA-SAB-EC-02-010. Office of the Administrator, Science Advisory Board, U.S. Environmental Protection Agency, Washington, DC. September 2002 [online]. Available: http://www.epa.gov/sab/pdf/ec02010.pdf [accessed Nov. 8, 2007].

EPASAB (U.S. Environmental Protection Agency Science Advisory Board). 2007. U.S. Environmental Protection Agency Chartered Science Advisory Board FY 2007 Member Biosketches. Science Advisory Board, U.S. Environmental Protection Agency, Washington, DC [online]. Available: http://www.epa.gov/sab/pdf/board_biosketches_2007.pdf [accessed Nov. 8, 2007].

Inside EPA's Risk Policy Report. 2007. Improved OMB Rating May Help Funding for EPA Ecological Research. Inside EPA's Risk Policy Report 14(39). September 25, 2007.

Morgan, G.M. 2007. Testimony M. Granger Morgan, Chair U.S. Environmental Protection Agency Science Advisory Board Before the Subcommittee on Energy and Environment Committee on Science and Technology, U.S. House of Representatives, March 14, 2007 [online]. Available: http://www.epa.gov/sab/pdf/2gm_final_written_testimony_03-14-07.pdf [accessed Nov. 8, 2007].

NRC (National Research Council). 1999. Evaluating Federal Research Programs: Research and the Government Performance and Results Act. Washington, DC: National Academy Press.

NRC (National Research Council). 2000. P. 99 in Strengthening Science at the U.S. Environmental Protection Agency: Research-Management and Peer-Review Practices. Washington, DC: The National Academy Press.

NRC (National Research Council). 2003. P. 3 in The Measure of STAR: Review of the U.S. Environmental Protection Agency's Science to Achieve Results (STAR) Research Grants Program. Washington DC: The National Academies Press.

NRC (National Research Council). 2007. Framework for the Review of Research Programs of the National Institute for Occupational Safety and Health. Aug. 10, 2007.

OMB (Office of Management and Budget). 2006a. Guide to the Program Assessment Rating Tool (PART). Office of Management and Budget. March 2006 [online].

Available: http://www.whitehouse.gov/omb/part/fy2006/2006_guidance_final.pdf [accessed Nov. 7, 2007].

OMB (Office of Management and Budget). 2006b. PART Appeals Board Decisions-Environmental Protection Agency. Letter from Clay Johnson, III, Deputy Director for Management, Office of Management and Budget, Washington, DC, to Marcus C. Peacock, Deputy Administrator, U.S. Environmental Protection Agency, Washington, DC. August 28, 2006.

OMB (Office of Management and Budget). 2007a. Guide to the Program Assessment Rating Tool (PART). Office of Management and Budget. January 2007 [online]. Available: http://stinet.dtic.mil/cgi-bin/GetTRDoc?AD=ADA471562&Location= U2&doc=GetTRDoc.pdf [accessed Nov. 7, 2007].

OMB (Office of Management and Budget). 2007b. Circular A–11, Part 7, Planning, Budgeting, Acquisition and Management of Capital Assets. Office of Management and Budget, Executive Office of the President. July 2007 [online]. Available: http://www.whitehouse.gov/omb/circulars/a11/current_year/part7.pdf [accessed Nov. 9, 2007].

OMB (Office of Management and Budget). 2007c. ExpectMore.gov. Office of Management and Budget [online]. Available: http://www.whitehouse.gov/omb/expect more/ [accessed Nov. 7, 2007].

Radin, B. 2006. Challenging the Performance Movement: Accountability, Complexity and Democratic Values. Washington, DC: Georgetown University Press.

Sarvana, A. 2007. Science Board Joins House in Opposing EPA Ecosystem Research Cuts. Inside EPA's Risk Policy Report 14 (37):1, 8-9. September 11, 2007.

Sinclair, R. 2007. Presentation at the Workshop on Evaluating the Efficiency of Research and Development Programs at the Environmental Protection Agency, April 24, 2007, Washington, DC.

3

Are the Efficiency Metrics Used by Federal Research and Development Programs Sufficient and Outcome-Based?

The second question the committee was asked to address is whether the efficiency metrics adopted by various federal agencies for R&D programs are "sufficient" and "outcome-based." The committee spent considerable time on this question, attempting to clarify why the issue has created such difficulties for agencies.

ATTEMPTING TO EVALUATE EFFICIENCY IN TERMS OF ULTIMATE OUTCOMES

In its guidance for undertaking Program Assessment Rating Tool (PART) evaluations, the Office of Management and Budget (OMB) clearly prefers that evaluation techniques should be related to the "outcomes" of the program; in other words, the metrics are to be outcome-based, including those for programs that perform R&D. For example, the R&D Investment Criteria state: "R&D programs should maintain a set of high priority, multi-year R&D objectives with annual performance outputs and milestones that show how one or more outcomes will be reached" (OMB 2007, p. 75).

In the case of the Environmental Protection Agency (EPA), that would mean that the efficiency of the research should be evaluated in terms of how much it contributes to improvements in the mission objectives of human health and environmental quality. However, at the same time as OMB presses for use of outcome metrics, it describes substantial difficulties in doing so. OMB points out, for example, that when the ultimate outcome of a research program is lives saved or avoidance of property damage, it may be the product of local or state actions, including political and regulatory actions that are beyond the agency's control and distant in time from the original research.

The committee believes that a link with ultimate outcomes is not the correct criterion for determining the sufficiency of metrics for evaluating research. Indeed, after analyzing agencies' attempts to measure outcome-based efficiency, the committee concluded that for most research programs ultimate-outcome-based metrics for evaluating the efficiency of research are *neither achievable nor valid*. The committee considers this issue to be of such importance for its report that it amplifies its reasoning as follows:

- There is often a large gap in time between the completion of research and the ultimate "outcome" of the research. In the case of EPA, for instance, the gap often is measured in years or even decades, commonly because the true outcome can be identified only by epidemiologic or ecologic studies that necessarily lag the original research itself. Thus, a retrospective outcome-based evaluation may be attempting to evaluate the "efficiency" of research conducted decades previously. Such an evaluation, if it can be done, may have little relevance to research being undertaken at the time of the evaluation.

- A number of entities over which the research program has no control are responsible for translating research results into outcomes. In the case of EPA, such translation can involve multiple steps even for problem-driven research. The EPA program office has to convert research results into a risk-management strategy that complies with legislative requirements. That strategy undergoes substantial review and comment by other government agencies, the regulated community, and the public before it can be adopted. It may even be subjected to judicial review. When it is finally adopted, state agencies usually perform the implementation chores with their own corresponding risk-management strategies and programs. Even then, no ultimate outcomes appear until people, businesses, or other government units take action in response to the programs and their accompanying rules and incentives. The initial research program has no influence over any of those steps. If the initial activity is core research, the number and variety of organizations and individuals involved in producing outcomes may be even greater.

- The results of research may change the nature of the outcome. The purpose of research is to produce knowledge, and new knowledge adds to the understanding of which outcomes are possible and desirable. To take another example of problem-driven research supported by EPA, suppose that results of a research project suggest that a particular chemical is toxic. That information may be only indicative, not definitive. EPA may launch a research program to confirm whether the chemical has toxic effects in humans or the natural environment. Results confirming toxicity would be expected to lead to a risk-management strategy that produces an ultimate outcome of reduced risk and improved health. In addition, EPA's research on chemicals and development of toxicity screening tests provide industry with tools that impact their choice of which chemicals to develop for the market. If the research provides no evidence of toxicity, no risk-management strategy will be developed; that is, there will be no "ultimate outcome." That would not mean that the research had no value; the "intermediate outcome" produced by the research would have provided knowledge that prevents unnecessary (and inefficient) actions, and the research would have been effective even though it did not produce reviewable ultimate outcomes.

Thus, ultimate outcomes of research are not useful criteria for measuring research efficiency, and ultimate-outcome-based metrics proposed by federal agencies to evaluate research efficiency cannot be sufficient.

PLACING "RESEARCH EFFICIENCY" IN PERSPECTIVE

If evaluating the efficiency of a research program in terms of ultimate outcomes is not feasible, what is the most appropriate (or "sufficient") way to evaluate a research program? Of first priority, in the committee's view, is an evaluation that is comprehensive—that applies all three categories of criteria used by PART (see Appendix G) and discussed above:

- *Relevance*—how well the research supports the agency's mission, including the timeliness of the project or program.
- *Quality*—the contribution of the research to expanding knowledge in a field and some attributes that define sound research in any context: its soundness, accuracy, novelty, and reproducibility.
- *Performance*—described in terms of both effectiveness, meaning the ability to achieve useful results,[1] and efficiency, meaning the ability to achieve quality, relevance, and effectiveness in timely fashion with little waste.

Nonetheless, OMB representatives stated at the workshop that a program is unlikely to receive a favorable review without a positive efficiency grade even if quality, relevance, and effectiveness are demonstrated. The committee rejected that approach because efficiency should not be evaluated independently but should be regarded as a relatively minor element of a comprehensive evaluation. As proposed in the PART guidance, efficiency constitutes only a portion of the performance criterion, which itself is one of the three major evaluation criteria.

Efficiency, of course, is a desirable goal, and it should be measured to the extent possible. That is commonly the case with input and output functions, whose efficiency may be clearly reflected in quantitative terms of hours, personnel, dollars, or other standard metrics. But undue emphasis on the single crite-

---

[1]That is, research of quality, relevance, and efficiency is effective only if the information it produces is in usable form.

rion of efficiency leads to imbalances. For example, if the main objective of a program is to receive, review, and return the grant proposals of researchers, it is desirable to do so promptly; but no one would recommend reducing review time to zero, because that practice, although undoubtedly efficient, would inevitably reduce the quality of review.

Many have cautioned against the reliance solely on quantifiable metrics for making decisions about the value of a program. This is particularly true if these metrics are collected solely to evaluate efficiency, rather than assessing the larger issue of the program's quality, relevance, and performance. Organizations often tend to manage what they measure, which can result in distortions in organizational emphasis and compromise the objectives of the program in question (Blau 1963). For example, in an assessment of the influence of the federal government's statistical requirements to evaluate local programs, De Neufville (1987, p. 346) found that "the required statistics seldom genuinely informed or directly affected program decisions by what they showed. Rather, they were assemblages of numbers tacked onto proposals. Indeed the preparation of the statistics was often delegated to a junior staff member or data analyst and done independently of the rest of planning...Local governments had little incentive or occasion to use them in any analytic way...Thus the required statistics became merely window dressing—part of the ritual of grant getting. As such they were not particularly accurate, but they were accepted. Few bothered to point out their limitations. It simply did not matter." Similarly, Weiss and Gruber's (1987) review of federal collection and distribution of education statistics noted that these can be guided by political influences.

PROCESS EFFICIENCY AND INVESTMENT EFFICIENCY

In the committee's view, the situation described in the preceding sections presents a conundrum, as follows:

- Demonstration of outcome-based efficiency of research programs is strongly urged for PART compliance.
- Ultimate-outcome-based metrics of research efficiency are neither achievable nor valid.

In the face of that conundrum, the committee found that PART asks two kinds of questions about efficiency—one explicit and one implicit. The explicit question applies to *inputs* and *outputs,* which should be identified and measured in their own right. Many cases of such efficiency can be characterized by the term *process efficiency.* That is most easily seen in aspects of R&D related to administration, facilities, and construction. Process efficiency can be measured by, for example, how fast a building is constructed, how closely a construction process adheres to budget, and what percentage of external grants is evaluated by peer review within a given period. Such activities can usually be described quantitatively (for example, in dollars or units of time) and measured against milestones, as described earlier in the case of earned-value management (EVM).

The implicit question about efficiency has to do with *ultimate outcomes*, for which PART prefers quantitative measures against milestones. In contrast to process activities, some major aspects of a research program cannot be evaluated in quantitative terms or against milestones.[2] The committee describes such aspects under the heading *investment efficiency,* the efficiency with which a research program is planned, funded, adjusted, and evaluated. Investment efficiency focuses on portfolio management, including the need to identify the most promising lines of research for achieving desired ultimate outcomes. It can be evaluated in part by assessing a program's strategic planning architecture. When an agency or research manager "invests" in research, the first step is to identify a desired outcome and a strategy to reach it. In general, investing in a research program involves close attention to three questions: are the right investments being made, is the research being performed at a high level of quality, and are timely and effective adjustments made in the multi-year course of the work to reflect new scientific information, new methods, and altered priorities? Those questions, especially the first, cannot be answered quantitatively, because the answers require judgment based on experience. Judgment is required to ensure that investment decisions are linked to strategic and multi-year plans (relevance), that the research is carried out at the highest level by the best people (quality), that funds are invested wisely in the right lines of research (effectiveness), and that the most economical management techniques are used to perform the research (efficiency). It is important to emphasize the value of ultimate outcomes for assessing the relevance of research. The concept of investment efficiency may be applied to studies that guide the next set of research projects or stepwise development of analytic tools or other products (Boer 2002).[3] An appropriate way to evaluate investment efficiency is to use expert-review panels, as described in Chapter 2. These considerations are indeed addressed by various PART questions in sections 1 and 2. However, the concept of investment efficiency, central to the performance of research, is not addressed in questions 3.4 or 4.3, those that specifically deal with efficiency.

WHAT ARE "SUFFICIENT" METRICS OF PROCESS EFFICIENCY?

As indicated in Chapter 2, no efficiency metrics currently used by agencies and approved by OMB to comply with PART succeed in measuring investment efficiency. Instead, the metrics address issues that fall under the heading of

---

[2]In its guidance, OMB recognizes the problem, concluding that "it may be difficult to express efficiency measures in terms of outcome. In such cases, acceptable efficiency measures could focus on how to produce a given output level with fewer resources" (OMB 2006, p. 10).

[3]Boer proposes a method for valuing plans by which the value may be analyzed quantitatively and increased by good management over time.

process efficiency. Process-efficiency metrics should meet the test of "sufficiency," however, and several questions can help in the framing of such a test. As the committee was tasked with evaluating whether these efficiency measures were "sufficient," it has developed its own questions for evaluating sufficiency below.

First, does the metric cover a representative portion of the program's operations? Metrics that pertain to only a small part of a program fail to indicate convincingly whether the program as a whole is managed efficiently. They may also create misguided incentives for program managers to improve the small portions of the program being assessed to the detriment of the rest. It is possible to use different metrics for different parts of a program, and indeed this approach may be useful to research-program managers; however, it is difficult to combine different metrics into a single number that represents the efficiency of an entire program. EPA acknowledged this difficulty when it noted that a major problem in applying any single metric across even a single agency is the variations among programs (see Appendix B).

Second, does the metric address both outputs and inputs of the program? The goal of a research program should be to produce desired outputs quickly and at minimal cost—that is, with minimal inputs—without diminishing their quality. Thus, a metric of efficiency should measure whether a program is producing its intended outputs. However, the use of such a metric requires that the program have some quality-assurance and quality-control (QA/QC) process to ensure that the drive for increased efficiency does not diminish the quality of the outputs.

Two other questions, although they do not directly determine sufficiency, should be asked about a proposed metric. The first is whether its use is likely to create undesirable incentives for researchers and research managers. A common adage in program evaluation is that "you get what you measure." A measurement system should not set up incentives that are detrimental to the operation of the program being evaluated (Drickhamer 2004). For example, program managers can often make adjustments that "meet the measure" without actually improving—if not adversely affecting—the program being evaluated.[4] Grizzle (2002) discussed the unintended consequences of the pervasive practice of performance measurement. She notes "we expect that measuring efficiency leads to greater efficiency and measuring outcomes leads to better outcomes, but we don't always get the results we expect."

---

[4]Drickhamer (2004) includes an example provided by Andy Carlino, a management consultant: "Carlino says he once worked for a large organization where plant managers received a bonus if they reduced direct labor. The easiest way to do that is to automate, which is what happened at this company in a big way. The result, he recalls, was more downtime and poor quality, which required more support by indirect labor, which led directly to customer quality and delivery issues. At the end of the day, direct labor went down, but total costs increased."

A second question—appropriate for all levels of compliance—is whether collecting the required information adds sizable administrative costs. Evaluation requirements, particularly when not carefully attuned to the program being evaluated, can cause program managers, administrators, and ultimately taxpayers substantial expense in collecting and processing information; the committee heard statements to that effect from research-intensive agencies attempting to comply with PART. Under the best of circumstances, the evaluation metric should depend on data already being collected to manage the program effectively and efficiently. Otherwise, the effort to measure the program's efficiency can reduce the efficiency desired by both the agency and OMB.

A CRITIQUE OF THE EFFICIENCY METRICS USED BY FEDERAL RESEARCH PROGRAMS

In light of those questions, it is appropriate to ask how well the metrics used by federal research programs meet the test of sufficiency. Chapter 2 described types of efficiency metrics that have been proposed or adopted by federal agencies to comply with PART (see Appendix E for details). The committee has examined nine of those metrics in the context of the four questions posed above and produced the following assessment.

1. Time to Process, Review, and Award Grants

Several agencies use time required to process grant requests as an efficiency metric. Such a metric is valuable if the awarding of grants is the purpose and primary output of the research program. In such cases, it would also satisfy the criterion of covering a substantial portion of the research program. One weakness, however, is that the measurement unit of inputs is time rather than a more inclusive metric of resources, such as total applicable administrative costs. As a result, changes in the program's administrative budget can result in changes in the metric that do not truly reflect changes in efficiency.[5] A second problem is that continual "improvement" of this metric—reducing the time required—meets a point of diminishing returns. Some amount of time is required to conduct efficient peer review and otherwise identify the best research proposals. Excessively reducing the resources (including time) devoted to those efforts could substantially reduce the quality, relevance, effectiveness, and even efficiency of the research being funded.

---

[5]For example, if a 20% decrease in administrative budget resulted in a 10% increase in the time required to award grants, this would appear as a decrease in efficiency according to the metric although it would be measured as an increase in efficiency if total costs were used as the denominator.

2. Proportion of Research Budget Consumed by Administrative Functions (Overhead Ratio)

This metric has the advantage of being able to cover the entire research program. A disadvantage is that it includes no measurement of research outputs. It cannot be improved indefinitely; some amount of administrative cost is necessary for managing a high-quality, relevant, effective, and, indeed, efficient research program. If administrative costs are reduced too far, all those characteristics will be lessened.

3. Publications Per Full-Time Equivalent or Per Dollar

Some agencies have proposed an efficiency metric of publications per FTE or per dollar. It is a useful metric for programs whose primary purpose is to produce publications, because it considers both inputs (FTEs or dollars) and outputs (publications). Here again some mechanism is needed to evaluate the relevance, quality, and effectiveness of the publications. Using a dollar metric rather than an FTE metric provides a better indication of efficiency because it considers total resources.

Some agencies use the number of peer-reviewed publications as a measure of program quality. A problem with this practice is that publication rates vary substantially among scientific disciplines (Geisler 2000). Therefore, using such a metric for a program that supports research in different disciplines can provide misleading results unless the publication rates in each discipline are normalized—for instance, by relating them to the mean rate for the discipline.[6]

4. Percentage of Work That Is Peer-Reviewed

Some agencies are apparently in the process of adopting a way to measure the portion of the research program that has been subjected to peer review as a metric of efficiency. Peer review is normally used to ensure the quality of research, not its efficiency. Asking experts who are experienced in managing research programs to evaluate the programs is an appropriate way to improve efficiency, but it is not clear that the agencies adopting this metric include such people in their review panels.

5. Average Cost Per Measurement or Analysis

Using such a metric might be sufficient for an agency if a large component of the "research" program is devoted to fairly repetitive or recurrent operations, such as analyzing the constituents of geologic or water samples, and adequate

---

[6]See NRC 2003 for further discussion of the problems involved in using such bibliometric analyses.

QA/QC procedures are incorporated into the analytic efforts. An advantage of this metric is that it considers both outputs and total costs. However, repetitive analyses or measurements are not normally a major component of an agency's research program, so they are unlikely to yield results that apply to a research effort as a whole.

6. Speed of Response to Information Requests

Several agencies use some metric of time required to respond to information requests. The technique suffers from the same weaknesses as the metric of time to respond to research requests and is even more likely to result in diminished quality because a QA/QC function, such as consumer satisfaction, is rarely incorporated into such programs to measure the quality of responses. Responding to information requests does not account for a substantial portion of an agency's research budget, so it is unlikely to measure the efficiency of a substantial portion of its research program.

7. Cost-Sharing

In a few instances, agencies are apparently using cost-sharing as a metric of efficiency. Cost-sharing may be a proxy for the quality, relevance, or effectiveness of some research to the extent that other agencies or private entities may be willing to share the costs of research that they consider to be of high quality, relevance, or effectiveness for their own missions. Cost-sharing does reduce the cost of research to the sponsoring agency, but it is not a metric of efficiency in itself; an increase in cost-sharing does not reduce the total resources devoted to research. It also fails to address research outputs.

8. Quality or Cost of Equipment and Other Inputs

Several agencies have adopted metrics that are related to the cost or efficiency of inputs to the research effort rather than to the research itself. Obviously, lower cost of inputs can result in lower-cost and therefore more efficient research—as long as the quality of the inputs does not correlate positively with their cost. Indeed, improved equipment (for example, more powerful computers or more advanced analytic equipment) can be important in improving the efficiency of research. However, this metric itself does not measure research efficiency. It does not consider outputs, and it focuses on only some inputs.

9. Variance from Schedule and Cost

EPA has recently agreed to attempt to use a variation of EVM to measure research efficiency. EVM measures the degree to which research outputs conform to scheduled costs along a timeline, but in itself it measures neither value nor efficiency. It produces a quantitative metric of adherence to a schedule and a budget. If the schedule or budget is inefficient, it is measuring how well the program adheres to an inefficient process.

This metric relies for its legitimacy on careful research planning and advanced understanding of the research to be conducted.[7] If a process generates a good research plan with distinct outputs (or milestones) produced in an efficient manner, measuring the agency's success in adhering to its schedule does become a metric of the efficiency of research management. It also satisfies the criteria of including outputs and applying to the entire research effort. Thus, it can be a sufficient metric of process efficiency so long as the underlying planning process incorporates the criteria of quality, relevance, and performance.

On the basis of those evaluations, the committee concludes that there may be some utility in certain proposed metrics for evaluating the process efficiency of research programs, particularly reduction in time or cost, on the basis of milestones, and reduction in overhead rate.

There may also be applications for some metrics in certain EPA research programs, including: reduction in average cost per measurement or analysis, adjusted for needed improvements; reduction in time or cost of site assessments; and reduction in time to process, review, and award extramural grants.

In all cases, caution should be used in applying the metrics, as consideration must be given to the type of program being evaluated.

FACTORS THAT REDUCE THE EFFICIENCY OF RESEARCH

Many forces outside the control of the researcher, the research manager, or OMB can reduce the efficiency of research, often in unexpected ways. Because these other forces can appreciably reduce the value of efficiency as a criterion by which to measure the results or operation of a research program, they are relevant here. For example,

- The efficiency of a research program is almost always adversely affected by reductions in funding. A program is designed in anticipation of a funding schedule. If funding is reduced after substantial funds are spent but before results are obtained, activities cannot be completed, and outputs will be lower than planned.
- When personnel ceilings are lowered, research agencies must hire contractors for research, and this is generally more expensive than in-house research.
- Infrastructure support consumes a large portion of the EPA Office of Research and Development (ORD) budget. Because the size and number of

---

[7]The method comes from the construction industry, in which scheduling and expected costs are understood better than they are in research.

laboratories and other entities are often controlled by political forces outside the agency, ORD may be unable to manage infrastructure efficiently.

• Inefficiencies may be introduced when large portions of the budget are consumed by congressional earmarks. That almost always constitutes a budget reduction because the earmarks are taken out of the budget that the agency had intended to use to support its strategic and multi-year plans at a particular level.

Still other factors may confound attempts to achieve and evaluate efficiency by formal, quantitative means. For example, the most efficient strategy in some situations is to spend more money, not less; a familiar example is the purchase of more expensive, faster computers. Or a research program may begin a search for an environmental hazard with little clue about its identity, and by luck a scientist may discover the compound immediately; does this raise the program's efficiency? Such examples seem to support the argument that an experienced and flexible research manager who makes use of quantitative tools as appropriate is the best "mechanism" for efficiently producing new knowledge, products, or techniques.

EVALUATING RESEARCH EFFICIENCY IN INDUSTRY

The committee reviewed information from the Industrial Research Institute (IRI), an association of companies established to enhance technical innovation in industry. IRI has been actively involved in developing metrics for use in evaluating R&D, studying the measurement of R&D effectiveness and productivity, and devising a menu of 33 metrics to be used in evaluating the effectiveness and productivity of R&D activities (Tipping et al. 1995).

A recent industry study surveyed 90 companies with revenues exceeding $1 billion in order to understand their efforts to monitor and manage the performance of their R&D activities. Preliminary evidence indicates that higher-performing companies differ from lower performing companies in the metrics they use to evaluate their R&D programs. For instance, higher performing companies are more likely to track pipeline productivity metrics (such as revenues from new products and the value of the portfolio in the pipeline) and portfolio health metrics (such as percentage of portfolio in short-, medium-, and long-term projects) than lower performing companies which are less likely to focus on business outcome metrics (such as margin growth or incremental market share) and typically use more metrics to manage the R&D organization (D. Garettson, RTEC, personal communication, December 7, 2007).

Other studies published in the technology-management literature conclude that efficiency is best evaluated secondarily to effectiveness. For example, Schumann et al. (1995) noted that the "key to efficiency is maximizing the use of internal resources, minimizing the time it takes to develop the technology, and maximizing the knowledge about the technology that product developers have readily available." However, they added that "rather than focus on efficiency, the focus of quality in R&D should be on effectiveness. The leverage here is ten to hundreds of times the R&D costs. . . . After the R&D organization has developed effectiveness, it can turn its attention to efficiency. The result must not be either/or, but rather simultaneous effectiveness and efficiency; i.e., doing the right things rightly."

THE SHORTCOMINGS OF RETROSPECTIVE REVIEW

Finally, PART calls for retrospective review of research programs. The most recent R&D Investment Criteria (OMB 2007, p. 72) state, "Retrospective review of whether investments were well-directed, efficient, and productive is essential for validating program design and instilling confidence that future investments will be wisely invested." Although periodic retrospective reviews for relevance are appropriate, as suggested in the Criteria (p. 74), the retrospective analysis can be an unreliable indicator of quality (recommended every 3-5 years; p. 75) and performance (recommended annually; p. 76), for several reasons:

- The size of any research program varies each year according to budget, so the amount and kind of work done also varies. Year-to-year comparisons can be invalid unless there is a constant (inflation-adjusted) stream of funding.
- Retrospective analysis cannot demonstrate that resources might not have been put to more productive uses elsewhere.
- Retrospective analyses are unlikely to influence future investment decisions, because they focus on expenditures in the past, when conditions and personnel were probably different.

Again, the value of any analysis depends on the experience and perspective of those who perform it and of those who integrate the results with other information to make research decisions.

SUMMARY AND RECOMMENDATIONS

Despite the desire of OMB for agencies to use outcome-based metrics to evaluate research efficiency, no such metric has been demonstrated, so none can be "sufficient."

Meaningful evaluation of the efficiency of research programs at federal agencies can take two distinct approaches. First, the *inputs* and *outputs* of a program can be evaluated in the context of *process efficiency* by using quantitative metrics, such as dollars or hours. Process-efficiency metrics cannot be applied to ultimate outcomes, but they can and should be applied to such capital-intensive R&D activities as construction, facility operation, and administration.

Second, research effectiveness can be evaluated in the context of *investment efficiency* by using expert-review panels to consider the relevance, quality, and performance of research programs. Research investment includes the activities in which a research program is planned, funded, adjusted, and evaluated. Excellence in those activities is most likely to lead to desired *outcomes*. An expert panel should begin its evaluation by examining a program in terms of its relevance, quality, and effectiveness, including how well the research is appropriate to strategic and multi-year plans. Once a panel has evaluated relevance, quality, and effectiveness, it is well positioned to judge how efficiently research is carried out.

The committee concludes that most evaluation metrics applied by federal agencies to R&D programs have been neither outcome-based nor sufficient. They have not been outcome-based, because ultimate-outcome-based efficiency metrics for research programs are *neither achievable nor valid.* Among the reasons is that ultimate outcomes are often removed in time from the research itself and may be influenced and even generated by entities beyond the control of the research program. They have not been sufficient, because most evaluation metrics purporting to measure process efficiency do not evaluate an entire program, do not evaluate the research itself, or fall short for other reasons explained in connection with the nine metrics evaluated above.

The use of inappropriate metrics to evaluate research can have adverse effects on agency performance and reputation. At the least, inappropriate metrics can provide an erroneous evaluation of performance at a considerable cost in data collection and analysis, not to mention disputes and appeals. At worst, program managers might alter their planning or management primarily to seek favorable PART ratings and thus compromise the results of research programs and ultimately weaken their outcomes. Agencies and oversight bodies alike should regard the evaluation of efficiency as a relatively minor part of the evaluation of research programs.

REFERENCES

Blau, P.M. 1963. The Dynamics of Bureaucracy: A Study of Interpersonal Relationships in Two Government Agencies. Chicago: University of Chicago Press.

Boer, F.P. 2002. The Real Options Solution: Finding Total Value in a High-Risk World, 1st Ed. New York: John Wiley & Sons.

de Neufville, J.I. 1987. Federal statistics in local governments. Pp. 343-362 in The Politics of Numbers, W. Alonso, and P. Starr, eds. New York: Russell Sage.

Drickhamer, D. 2004. You get what you measure. Ind. Week, Aug. 1, 2004 [online]. Available: http://www.industryweek.com/CurrentArticles/Asp/articles.asp?Article Id=1658 [accessed Nov. 9, 2007].

Geisler, E. 2000. The Metrics of Science and Technology. Westport, CT: Quorum Books.

Grizzle, G. 2002. Performance measurement and dysfunction: The dark side of quantifying work. Public Perform. Manage. Rev. 25(4):363-369.

NRC (National Research Council). 2003. The Measure of STAR: Review of the U.S. Environmental Protection Agency's Science to Achieve Results (STAR) Research Grants Program. Washington, DC: The National Academies Press.

OMB (Office of Management and Budget). 2006. Guide to the Program Assessment Rating Tool (PART). Office of Management and Budget. March 2006 [online]. Available: http://www.whitehouse.gov/omb/part/fy2006/2006_guidance_final.pdf [accessed Nov. 7, 2007].

OMB (Office of Management and Budget). 2007. Guide to the Program Assessment Rating Tool (PART). Office of Management and Budget. January 2007 [online]. Available: http://stinet.dtic.mil/cgibin/GetTRDoc?AD=ADA471562&Location= U2&doc=GetTRDoc.pdf [accessed Nov. 7, 2007].

Schumann, P.A., D.L. Ransley, and C.L. Prestwood. 1995. Measuring R&D performance. Res. Technol. Manage. 38(3):45-54.

Tipping, J.W., E. Zeffren, and A.R. Fusfeld. 1995. Assessing the value of your technology. Res. Technol. Manage. 38(5):22-39.

Weiss, J.A. and J.I. Gruber. 1987. The managed irrelevance of federal education statistics. Pp. 363-391 in The Politics of Numbers, W. Alonso, and P. Starr, eds. New York: Russell Sage.

4

A Model for Evaluating Research and Development Programs

This report has discussed the difficulty of evaluating research programs in terms of results, which are usually described as outputs and ultimate outcomes. However, between outputs and ultimate outcomes are many kinds of "intermediate outcomes" that have their own value as results and can therefore be evaluated.

The following is a sample of the kinds of activities that might be categorized as outputs, intermediate outcomes, and ultimate outcomes:

- *Outputs* include peer-reviewed publications, databases, tools, and methods.
- *Intermediate outcomes* include an improved body of knowledge available for decision-making, integrated science assessments (previously called criteria documents), and the dissemination of newly developed tools and models.
- *Ultimate outcomes* include improved air or water quality, reduced exposure to hazards, restoration of wetland habitats, cleanup of contaminated sediments, and demonstrable improvements in human health.

Those steps can be described in different terms, depending on the agency using them and the scope of the research involved. For the Environmental Protection Agency (EPA) Office of Research and Development (ORD), for example, results that might fit the category of intermediate outcome might be: the provision of a body of knowledge that can be used by EPA's customers and the use of that knowledge in planning, management, framing of environmental regulations, and other activities. Intermediate outcomes are bounded on one side by outputs (such as toxicology studies, reports of all kinds, models, and monitoring activities) and on the other side by ultimate outcomes (such as protection and improvement of human health and ecosystems).

As a somewhat idealized example of how EPA (or other agencies) might conceptualize and make use of these terms, the following logic model shows the sequence of research, including inputs, outputs, intermediate outcomes, and ultimate outcomes. These stages in the model are roughly aligned with various events and users as research knowledge is developed. However, it is important to recognize that this model must be flexible to respond to rapid changes in research direction based upon unanticipated issues. The shift of personnel and resources to meet a new or newly perceived environmental challenge inevitably will impact the ability to complete planned R&D programs.

In the top row of Figure 4-1, the logic flow begins with process inputs and planning inputs. Process inputs could include budget, staff (including the training needed to keep a research program functioning effectively), and research facilities. Planning inputs could include stakeholder involvement, monitoring data, and peer review. Process and planning inputs are transformed into an array of research activities that generate research outputs listed in the first ellipse, such as recommendations, reports, and publications. The combination of research and research outputs leads to intermediate outcomes.

A helpful feature of the model is that there are two stages of intermediate outcomes: research outcomes and customer outcomes. The intermediate research outcomes are depicted in the arrow and include an improved body of knowledge available for decision-making, new tools and models disseminated, and knowledge ready for application. The intermediate research outcomes in the arrow are followed by intermediate customer outcomes, in the ellipse, that describe a usable body of knowledge, such as regulations, standards, and technologies. Intermediate customer outcomes also include education and training. They may grow out of integrated science assessments or out of information developed by researchers and help to transform the research outputs into eventual ultimate outcomes. The customers who play a role in the transformation include international, national, state, and local entities and tribes; nongovernment organizations; the scientific and technical communities; business and industry; first responders; decision-makers; and the general public. The customers take their own implementation actions, which are integrated with political, economic, and social forces.

The use of the category of intermediate outcome does not require substantial change in how EPA plans and evaluates its research. The strategic plan of ORD, for example, already defines the office's mission as to "conduct leading-edge research" and to "foster the sound use of science" (EPA 2001). Those lead naturally into two categories of intermediate outcome: intermediate outcomes from research and intermediate outcomes from users of research.

EPA's and ORD's strategic planning architecture fits into the logic diagram as follows: the ellipse under "Research Outputs" contains the annual performance metrics and the annual performance goals (EPA 2007b), the arrow under "Intermediate Outcomes from Research" contains sub-long-term goals, the ellipse under "Intermediate Outcomes from Users of Research" contains the

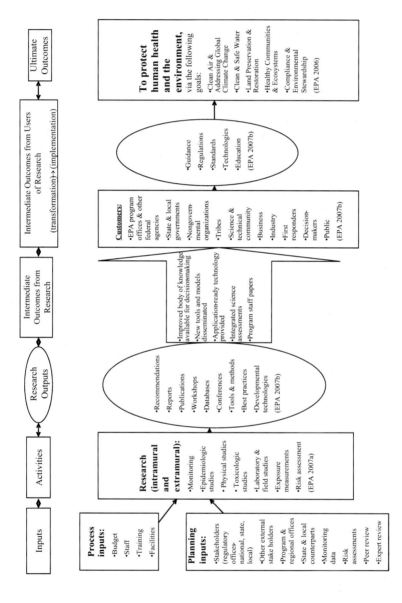

FIGURE 4-1 EPA research presented as a logic model. Source: Modified from NRC 2007.

long-term goals (EPA 2007b), and the box under "Ultimate Outcomes" contains EPA's overall mission (EPA 2006). In general, ultimate outcomes are evaluated at the level of the mission, intermediate outcomes at the level of multi-year plans, and outputs at the level of milestones.

Specific examples of outputs, intermediate outcomes, and ultimate outcomes taken from the Ecological Research Multi-Year plan (EPA 2003),[1] fit into the framework as follows:

- *Outputs*: a draft report on ecologic condition of western states, and the baseline ecologic condition of western streams determined.
- *Intermediate outcome from research*: a monitoring framework is available for streams and rivers in the western United States that can be used from the local to the national level for statistical assessments of condition and change.
- *Intermediate outcome from customers*: the states and tribes use a common monitoring design and appropriate ecologic indicators to determine the status and trends of ecologic resources.
- *Ultimate outcomes*: critical ecosystems are protected and restored (EPA objective), healthy communities and ecosystems are maintained (EPA goal), and human health and the environment are protected (EPA mission).

Similar logic models might be drawn from EPA's other multi-year plans, including water-quality monitoring and risk-assessment protocols for protecting children from pesticides.

The use of the model can have several benefits. First, it can help to generate understanding of whether and how specific programs transform the results of research into benefits for society. The benefits—for example, an identifiable improvement in human health—may take time to appear because they depend on events or trends beyond EPA's influence. The value of a logic model is to help to see important intermediate points in development that allow for evaluation and, when necessary, changes of course.

Second, the model can help to "bridge the gap" between outputs and ultimate outcomes. For a project that aims to improve human health through research, for example, there are too many steps and too much time between the research and the ultimate outcomes to permit annual evaluation of the progress or efficiency of a program. The use of intermediate outcomes can add results that are key steps in its progress.

The use of intermediate outcomes can also give a clearer view of the value of negative results. Such results might seem "ineffective and inefficient" to an evaluator, perhaps on the grounds that the project produced no useful practice or product. Making use of intermediate outcomes in the reviewing process, how-

---

[1]Note that p. 14 (EPA 2003) shows a logic diagram of how all the sub-long-term goals connect to feed into the long-term goal.

ever, may clarify that a negative result is actually "effective and efficient" if it prevents wasted effort by closing an unproductive line of pursuit.

Intermediate outcomes are already suggested by the section of the 2007 PART guidance entitled Categories of Performance Measures (OMB 2007, p. 9). The guidance acknowledges the difficulty of using ultimate outcomes to measure efficiency, and proposes the use of proxies when difficulties arise, as in the following example:

> Programs that cannot define a quantifiable outcome measure—such as programs that focus on process-oriented activities (e.g., data collection, administrative duties or survey work)—may adopt a "proxy" outcome measure. For example, the outcomes of a program that supplies forecasts through a tornado warning system could be the number of lives saved and property damage averted. However, given the difficulty of measuring those outcomes and the necessity of effectively warning people in time to react, prepare, and respond to save lives and property, the number of minutes between the tornado warning issuance and appearance of the tornado is an acceptable proxy outcome measure.

Identification of intermediate steps brings into the PART process an important family of existing results that may lend themselves to qualitative and sometimes quantitative assessment, which can provide useful new data points for reviewers. The terms in which those steps are described depend on the agency, its mission, and the nature and scope of its work.

SUMMARY

Although the task of reviewing research programs is complicated by the limitations of ultimate-outcome-based metrics, the committee suggests as a partial remedy the use of additional results that might be termed intermediate outcomes. This class of results, intermediate between outputs and ultimate outcomes, could enhance the evaluation process by adding trackable items and a larger body of knowledge for decision-making. The additional data points could make it easier for EPA and other agencies to see whether they are meeting the goals they have set for themselves, how well a program supports strategic and multi-year plans, and whether changes in course are appropriate. Using this class of results might also improve the ability to track progress annually.

REFERENCES

EPA (U.S. Environmental Protection Agency). 2001. Strategic Plan. EPA/600/R-01/003. Office of Research and Development, U.S. Environmental Protection Agency, Washington, DC. January 2001 [online]. Available: http://www.epa.gov/osp/strtplan/documents/final.pdf [accessed Nov. 13, 2007].

EPA (U.S. Environmental Protection Agency). 2003. Sub-long-term goals, annual performance goals and annual performance measures for each long term goal. Appendix 1 of the Ecological Research Multi-Year Plan. Office of Research and Development, U.S. Environmental Protection Agency. May 29, 2003 Final Version [online]. Available: http://www.epa.gov/osp/myp/eco.pdf [accessed Nov. 1, 2007].

EPA (U.S. Environmental Protection Agency). 2006. EPA Strategic Plan 2006-2011: Charting Our Course. U.S. Environmental Protection Agency. September 30, 2006 [online]. Available: http://www.epa.gov/cfo/plan/2006/entire_report.pdf [accessed Nov. 13, 2007].

EPA (U.S. Environmental Protection Agency). 2007a. Research Programs. Office of Research and Development, U.S. Environmental Protection Agency [online]. Available: http://www.epa.gov/ord/htm/researchstrategies.htm [accessed Nov. 13, 2007].

EPA (U.S. Environmental Protection Agency). 2007b. Research Directions: Multi-Years Plans. Office of Science Policy, U.S. Environmental Protection Agency [online]. Available: http://www.epa.gov/osp/myp.htm [accessed Nov. 13, 2007].

OMB (Office of Management and Budget). 2007. Guide to the Program Assessment Rating Tool (PART). Office of Management and Budget. January 2007 [online]. Available: http://stinet.dtic.mil/cgi-bin/GetTRDoc?AD=ADA471562&Location=U2&doc=GetTRDoc.pdf [accessed Nov. 7, 2007].

NRC (National Research Council). 2007. Framework for the Review of Research Programs of the National Institute for Occupational Safety and Health. Aug. 10, 2007.

5

Findings, Principles, and Recommendations

In this chapter, the committee draws together its preceding discussions in the form of findings, principles, and recommendations. The findings constitute a brief summary of major points discussed in Chapters 1-4. The principles are intended for use by both the Environmental Protection Agency (EPA) and other research-intensive federal agencies. The recommendations are intended specifically for EPA, although other agencies, including the Office of Management and Budget (OMB), may find them useful.[1]

To introduce this chapter, it is useful to begin with the two central issues on which the committee focused many of its discussions. The first is the emphasis on efficiency in Program Assessment Rating Tool (PART) reviews. In the planning, execution, and review of research programs, efficiency should normally be subordinate to the criteria of relevance, quality, and effectiveness for reasons explained in Chapter 3. However, all federal programs should use efficient spending practices, and the committee suggests which aspects of efficiency can be measured in research programs and how that might best be done. Two kinds of efficiency should be differentiated. The first, *process efficiency*, uses primarily quantitative metrics to evaluate management processes whose results are known and for which benchmarks can be defined and progress can be measured against milestones. The second, *investment efficiency*, measures how well a program's resources have been invested and how well they are being managed. Evaluating investment efficiency involves qualitative measures, primarily the judgment and experience of expert-review panels, and may also draw on quantitative data. Investment efficiency is the responsibility of the portfolio manager who identifies the most promising lines of research for achieving desired outcomes.

---

[1] It should be emphasized again that these recommendations apply only to R&D programs, not to the much broader universe of federal programs to which PART is applied.

The second central issue is the charge question of whether metrics used by federal agencies to measure the efficiency of research are "sufficient" and "outcome-based." In approaching sufficiency, the committee gathered examples of methods used by agencies and organized them in nine categories. It found that most of the methods were insufficient for evaluating programs' process efficiency either because they addressed only a portion of a program or because they addressed issues other than research, and all were insufficient for evaluating investment efficiency because they did not include the use of expert review.

In responding to the question of whether the metrics used are outcome-based, the committee determined that ultimate-outcome-based evaluations of the efficiency of research are *neither achievable nor valid*. The issue is discussed in Chapter 3.

Those two basic conclusions constitute the background of the major findings of this report. Findings 2, 4, 5, 6, and 7 are linked to specific charge questions, as indicated; findings 1 and 3 are more general.

FINDINGS

1. The key to research efficiency is good planning and implementation. EPA and its Office of Research and Development (ORD) have a sound strategic planning architecture that provides a multi-year basis for the annual assessment of progress and milestones for evaluating research programs, including their efficiency.

2. All the metrics examined by the committee that have been proposed by or accepted by OMB to evaluate the efficiency of federal research programs have been based on the inputs and outputs of research-management processes, not on their outcomes.

3. Ultimate-outcome-based efficiency metrics are neither achievable nor valid for this purpose.

4. EPA's difficulties in complying with PART questions about efficiency (questions 3.4 and 4.3[2]) have grown out of inappropriate OMB requirements for outcome-based efficiency metrics.

5. An "ineffective" (OMB 2007a)[3] PART rating of a research program can have serious adverse consequences for the program or the agency.

---

[2]Question 3.4 is "Does the program have procedures (e.g. competitive sourcing/cost comparisons, IT improvements, appropriate incentives) to measure and achieve efficiencies and cost effectiveness in program execution?" Question 4.3 is "Does the program demonstrate improved efficiencies or cost effectiveness in achieving program goals each year?"

[3]OMB (2007a) states that "programs receiving the Ineffective rating are not using tax dollars effectively. Ineffective programs have been unable to achieve results due to a lack of clarity regarding the program's purpose or goals, poor management, or some other significant weakness. Ineffective programs are categorized as Not Performing."

6. Among the metrics proposed to measure process efficiency, several can be recommended for wider use by agencies (see recommendation 1).

7. The most effective mechanism for evaluating the investment efficiency of R&D programs is an expert-review panel, as recommended in earlier reports of the Committee on Science, Engineering, and Public Policy and the Board on Environmental Studies and Toxicology. Expert-review panels are much broader than scientific peer-review panels.

PRINCIPLES

The foregoing findings led to a series of principles that the committee used to address the overall process of evaluating research programs in the context of agency long-term plans and missions. A central thesis of this report is that the evaluation principles can and should be applied to all federally supported research programs and can also be applied to research in other contexts. The committee hopes that these principles will be adopted by EPA and other research-intensive agencies in assessing their R&D programs.

Principle 1

Research programs supported by the federal government should be evaluated regularly to ensure the wise use of taxpayers' money.

The purpose of OMB's PART is to ensure that the government is spending taxpayers' money wisely. This committee's recommendations are designed to further that aim. More broadly, the committee agrees that the research programs of federal agencies should be evaluated regularly, as are other programs of the federal government.

During the evaluations, efforts should be made to evaluate the efficiency of the research programs of agencies. The development of tools for doing that is still in an early stage, and agencies continue to negotiate their practices internally and with OMB. EPA's multi-year plans, which provide an agency-wide structure to review progress and to revise annually, constitute a useful framework for organizing evaluations that serve as input into the PART process.

Principle 2

Despite the wide variability of research activities among agencies, all agencies should evaluate their research efforts according to the same criteria: relevance, quality, and performance.

Those criteria are defined in this report as follows:

- *Relevance* is a measure of how well research supports an agency's mission.
- *Quality* is a measure of the novelty, soundness, accuracy, and reproducibility of research.
- *Performance* is described in terms of both effectiveness (the ability to achieve useful results) and efficiency (the ability to achieve quality, relevance, and effectiveness in timely fashion and with little waste).

The research performed by federal agencies varies widely by primary mission responsibility. The missions of the largest research-intensive agencies include defense, energy, health, space, agriculture, and the environment. Their research efforts share assumptions, approaches, and investigative procedures, so they should be evaluated by the same criteria.

Research that is designed appropriately for a mission (relevance), is implemented in accordance with sound research principles (quality), and produces useful results (effectiveness) should be managed and performed as efficiently as possible. That is, research of unquestionable quality, relevance, and efficiency is *effective* only if the information it produces is in a usable form. The committee emphasizes that research effectiveness, in the context of PART, is achieved only to the degree that the program manager makes the most effective use of resources by allocating resources to the most appropriate lines of investigation. This integrated view is a reasonable starting point for the evaluation of research programs.

Principle 3

The process efficiency of research should not be evaluated using outcome-based metrics.

PART encourages the use of outcome-based metrics to evaluate the efficiency of federal programs. For many or perhaps most programs, especially those with clearly defined and predictable outcomes, such as countable services, that is an appropriate and practical approach that makes it possible to see how well inputs (resources) have been managed and applied to produce outputs. But OMB recognizes the difficulty of using outcome-based metrics to measure the efficiency of core or basic-research programs. According to PART guidance (OMB 2007b),

agencies should define appropriate output and outcome measures for all R&D programs, but agencies should not expect fundamental basic research to be able to identify outcomes and measure performance in the same way that applied research or development are able to. Highlighting the results of basic research is important, but it should not come at the expense of risk-taking and innovation. For some basic research programs,

OMB may accept the use of qualitative outcome measures and quantitative process metrics (OMB 2007b, p. 76).

The committee agrees with that view, as elaborated below, and finds that ultimate-outcome-based efficiency metrics are neither achievable nor valid, as explained in Chapter 3.

In some instances, however, it may be useful to reframe the results of research to include the category of *intermediate outcomes,* the subject of Chapter 4. That category of results may include new tools, models, and knowledge for use in decision-making. Because intermediate outcomes are available sooner than ultimate outcomes, they may provide more practical and accessible metrics for agencies, expert-review panels, and oversight bodies.

Principle 4

The efficiency of R&D programs can be evaluated on the basis of two metrics: investment efficiency and process efficiency.

In the committee's view, the construct presented by PART has proved unworkable for research-intensive agencies partly because of their difficulty in evaluating the "efficiency" of research. In lieu of that construct, the committee suggests that any evaluation of a research program be framed around two questions: Is the program making the right investments? Is it managing those investments well?

This report has used the term *investment efficiency* for the first evaluation metric. Investment efficiency is determined by examining a program in light of its relevance, quality, and performance—in other words, by asking whether the agency has invested in the right research portfolio and managed it wisely. Those criteria are most relevant to research *outcomes.*

The issue of efficiency is not the central concern in asking whether a program is making the right investments. But it is implicit in that the portfolio manager must make wise research investments if the program is to be effective and efficient; once resources, which are always finite, have been invested, they must be used to optimize results.

The totality of those activities might be called portfolio management, a more familiar term that suggests linking research activities with strategic and multi-year plans. Sound portfolio management is the surest route to desired outcomes.

The elements of investment efficiency are addressed in most agency procedures developed under the Government Performance and Results Act (GPRA) and in PART questions, although not in those addressing efficiency (that is, questions 3.4 and 4.3). Moreover, it is essential to correct the misunderstanding embodied in the following statement in the PART guidance: "Programs must document performance against previously defined output and outcome metrics"

(OMB 2007b, p. 76). A consistent theme of the present report is that for many research programs there can be no "outcome metrics"; that is true especially for core research, as discussed in Chapter 3.

Distinct from investment efficiency is *process efficiency*, which has to do with how well research investments are managed. Process efficiency involves activities whose results are well known in advance and can be tracked by using established benchmarks in such quantities as dollars and hours.

Process efficiency is secondary to investment efficiency in that it adds value only after a comprehensive evaluation of relevance, quality, and effectiveness. Process efficiency most commonly addresses *outputs*, which are the near-term results of research. It can also—like investment efficiency—make use of intermediate outcomes, which can be identified earlier than ultimate outcomes and thus provide valuable data points for reviewers.

Principle 5

Investment efficiency is best evaluated by expert-review panels that use primarily qualitative measures tied to long-term plans.

PART questions 3.4 and 4.3 seem to require evaluation of the efficiency of research in isolation from review of relevance and quality and thus emphasize cost and time. Agencies find that this approach may place programs at risk because the failure to satisfy PART on efficiency-related questions can increase the chances of an unacceptable rating for the total R&D program. As discussed in Chapter 3, quantitative metrics in the context of quality and relevance are important in measuring process efficiency but by themselves cannot assess the value of a research program or identify ways to improve it.

A more appropriate approach is to adapt the technique of expert review, already recommended by the National Research Council for compliance with GPRA. Indeed, OMB (2007b, p. 76) specifically recommends, in its written instructions to agencies, that agency managers "make the processes they use to satisfy the Government Performance and Results Act (GPRA) consistent with the goals and measures they use to satisfy these [PART] R&D criteria."

One advantage of using an expert-review panel is its ability to evaluate both investment efficiency and process efficiency. It can determine the kind of research that is most appropriate for advancing the mission of an agency and the best management strategies to optimize the results of the research with the resources available.

An expert-review panel can also identify emerging issues and their place in the research portfolio. Those would be developing fields (for example, nanotechnology a decade ago) identified by the agency for their potential importance but not mature enough for inclusion in a strategic plan. Identification of new fields might be thought of as an intermediate outcome because their value can be anticipated as a result of continuing core or problem-driven research and through the process of long-term planning. Because they may not seem urgent enough to have a place in a current strategic plan, emerging issues often fall victim to the budget-cutter's knife, even though an early start on a new topic can bring long-term efficiencies and strengthen research capabilities.

Principle 6

Process efficiency, which may be evaluated by using both expert review and quantitative metrics, should be treated as a minor component of research evaluation.

PART question 3.4, the one that addresses efficiency most explicitly, asks of every federal program whether it has procedures "to measure and achieve efficiencies and cost effectiveness in program execution", including "at least one efficiency measure that uses a baseline and targets" (EPA 2007b, p. 41). Research programs, especially programs of core or basic research, are unlikely to be able to respond "yes" to that question, because research managers cannot set baselines and targets for investigations whose outcomes are unknown. Therefore, such programs are unlikely to gain a "yes" for the question and are less likely to receive an acceptable rating under PART. In addition, failure on the PART efficiency questions precludes a "green" score on the Budget-Perfor mance Integration initiative of the President's Management Agenda.[4] Isolating efficiency as an evaluation criterion can produce a picture that is at best incomplete and at worst misleading. It is easy to see how an effort to reduce the time or money spent on a project, in order to increase efficiency, might also reduce its quality unless this effort is part of a comprehensive evaluation.

To evaluate applied research, especially in a regulatory agency, such as EPA, it is essential to understand the strategic and multi-year plans of the regulatory offices, the anticipated contributions of knowledge from research to plans and decisions, and the rather frequent modifications of plans due to intervening judicial, legislative, budgetary, or societal events and altered priorities. Some of those intervening events may be driven by new scientific findings.

The efficiency of research-management processes should certainly be evaluated. They include such activities as grant administration, facility maintenance or construction, and repeated events, such as air-quality sampling. Process efficiency can be evaluated with quantitative management tools, such as earned-value management (EVM). But such evaluations should be integrated with the work of expert-review panels if they are to contribute to the larger task of program evaluation.

---

[4]PART Guidance states, "The President's Management Agenda (PMA) Budget and Performance Integration (BPI) Initiative requires agencies to develop efficiency measures to achieve *Green* status" (OMB 2007b, p. 9).

In summary, efficiency measurements should not dominate or override the overall evaluation of a research program. Parts of the program may not be amenable to quantitative metrics, and the absence of quantitative metrics should not be cause for a low rating that harms the reputation of the program or the agency.

RECOMMENDATIONS

The following recommendations flow from the committee's conclusion that undue emphasis has been placed on the single criterion of efficiency. That emphasis, which is often seen for non-R&D activities throughout the main body of the PART instructions, is not explicit in the PART Investment Criteria (OMB 2007b). Rather, it has emerged during agency reviews, appeal rulings, and outside evaluations of the PART process, despite its inappropriateness for the evaluation of research programs. The issue is important because unsatisfactory responses to the two PART efficiency-focused questions have apparently contributed to a low rating for an entire program (for example, EPA's Ecological Research Program) and later budget cuts (Inside EPA's Risk Policy Report 2007).[5] Evaluation of research should begin not with efficiency but with the criteria of relevance, quality, and effectiveness and should secondarily address efficiency only after these criteria have been reviewed.

Recommendation 1

To comply with PART, EPA and other agencies should only apply quantitative efficiency metrics to measure the *process efficiency* of research programs. Process efficiency can be measured in terms of inputs, outputs, and some intermediate outcomes but not in terms of ultimate outcomes.

For compliance with PART, evaluation of the efficiency of a research program should not be based on ultimate outcomes. Ultimate outcomes can seldom be known until considerable time has passed after the conclusion of the research. Although PART documents encourage the use of outcome-based metrics, they also describe the difficulty of applying them.

Given that restriction, the committee recommends that OMB and other oversight bodies focus not on *investment efficiency* but on *process efficiencies* when addressing questions 3.4 and 4.3—the ways in which program managers exercise skill and prudence in conserving resources. For evaluating process efficiency, quantitative methods can be used by expert-review panels and others to track and review the use of resources in light of goals embedded in strategic and

---

[5]According to EPA's Risk Policy Report, "previous PART reviews criticized ERP [the Ecological Research Program] for not fully demonstrating the results of programmatic and research efforts – and resulted in ERP funding cuts." (Inside EPA's Risk Policy Report 2007)

multi-year plans. Moreover, to facilitate the evaluation process, the committee recommends including *intermediate outcomes*, as distinguished from *ultimate outcomes*. Intermediate outcomes include such results as an improved body of knowledge available for decision-making, comprehensive science assessments, and the dissemination of newly developed tools and models.

The PART R&D investment-criteria document (OMB 2007b, see also Appendix G) should be revised to make it explicit that quantitative efficiency metrics should be applied only to process efficiency.

Recommendation 2

EPA and other agencies should use expert-review panels to evaluate the *investment efficiency* of research programs. The process should begin by evaluating the relevance, quality, and performance[6] of the research.

OMB should make an exception when evaluating R&D programs under PART to permit evaluation of investment efficiency as well as process efficiency. This approach will make possible a more complete and useful evaluation.

Investment efficiency is used in this report to indicate whether an agency is "doing the right research and doing it well." The term is used as a gauge of portfolio management to measure whether a program manager is investing in research that is relevant to the agency's mission and long-term plans, whether the research is being performed at a high level of quality, and whether timely and effective adjustments are being made in the multi-year course of the work to reflect new scientific information, new methods, and altered priorities. Those questions cannot be answered quantitatively; they require judgment based on experience. The best mechanism for measuring investment efficiency is the expert-review panel. The concept of investment efficiency may be applied to studies that guide the next set of research projects and stepwise development of analytic tools or other products.

EPA should continue to obtain primary input for PART compliance by using expert review under the aegis of its Board of Scientific Counselors (BOSC) and Science Advisory Board (SAB). Expert review provides a forum for evaluation of research outcomes and complements the efforts of program managers in their effort to adjust research activities according to multi-year plans and anticipated outcomes. To enhance the process, consideration should be given to *intermediate outcomes*. As outputs and intermediate outcomes are achieved, the expert-review panel can use them to adjust and evaluate the expected ultimate outcomes (see Logic Model in Chapter 4).

---

[6]*Performance* is described in terms of both effectiveness (the ability to achieve useful results) and efficiency (the ability to achieve research quality, relevance, and effectiveness with little waste).

The qualitative emphasis of expert review should not obscure the importance of quantitative metrics, which should be used whenever possible by expert-review panels to evaluate process efficiency when activities can be measured quantitatively and linked to milestones—for example, administration, construction, grant administration, and facility operation.

In evaluating research at EPA, both EPA and OMB should place greater emphasis on identifying emerging and cross-cutting issues. ORD needs to be responsive to short-term R&D requests from the program offices, but it must have an organized process for identifying future research needs. BOSC and SAB should assign appropriate weight in their evaluations to forward-looking exercises that sustain the agency's place at the cutting edge of mission-relevant research.

Expert-review panels and oversight bodies should recognize that research managers need the flexibility to adapt to the realities of input changes beyond the agency's control, especially budgeting adjustments. The most rigorous planning cannot foresee the steps that might be required to maintain efficiency in the face of recurrent unanticipated change.

Recommendation 3

The efficiency of research programs at EPA should be evaluated according to the same overall standards used at other agencies.

EPA has failed to identify a means of evaluating the efficiency of its research programs that complies with PART to the satisfaction of OMB. Some of the metrics it has proposed, such as the number of publications per full-time equivalent (FTE), have been rejected, although accepted by OMB for other agencies. OMB has encouraged EPA to apply the common management technique of EVM, which measures the degree to which research outputs conform to scheduled costs along a timeline, but EPA has not found a way to apply EVM to research activities themselves. No other agency has been asked to use EVM for research activities, and none has done so.

Agencies have addressed PART questions with different approaches, which are often not in alignment with their long-term strategies or missions. Many of the approaches refer only to portions of programs, quantify activities that are not research activities, or review processes that are not central to R&D programs. In short, many federal agencies have addressed PART with responses that are not, in the wording of the charge, "sufficient."

ADDITIONAL RECOMMENDATION FOR THE OFFICE OF MANAGEMENT AND BUDGET

OMB should have oversight and training programs for budget examiners to ensure consistent and equitable implementation of PART in the many agencies that have substantial R&D programs.

Evaluating different agencies by different standards is undesirable because results are not comparable. OMB budget examiners bear primary responsibility for working with agencies in PART compliance and in interpreting PART questions for the agencies. Although not all examiners can be expected to bring scientific training to their discussions with program managers, they must bring an understanding of the research process as it is performed in the context of federal agencies, as discussed in Chapters 1-3.[7]

OMB decisions about whether to accept or reject metrics for evaluating the efficiency of research programs have been inconsistent. A decision to reject the metrics of one agency while accepting similar metrics at another agency can unfairly damage the reputation of the first agency and diminish the credibility of the evaluation process itself. Because the framework of PART is virtually the same for all agencies and because the principles of scientific inquiry are virtually the same in all disciplines, the implementation of PART should be both consistent and equitable in all federal research programs.

It should be noted that actual consistency is unlikely to be achieved in the vast and varied universe of government R&D programs, which fund extramural basic research, mission-driven intramural labs, basic research labs, construction projects, facilities operations, prototype development, and many other operations. Indeed, it is difficult even to define consistent approaches that would be helpful to both agencies and the OMB. But there is ample room for examiners to provide clearer, more explicit directions, understand the particular functioning of R&D programs, and discern cases when exceptions to broad requirements are appropriate.

REFERENCES

Inside EPA's Risk Policy Report. 2007. Improved OMB Rating May Help Funding for EPA Ecological Research. Inside EPA's Risk Policy Report 14(39). September 25, 2007.

OMB (Office of Management and Budget). 2007a. ExpectMore.gov. Office of Management and Budget [online]. Available: http://www.whitehouse.gov/omb/ expectmore/ [accessed Nov. 7, 2007].

OMB (Office of Management and Budget). 2007b. Program Assessment Rating Tool Guidance No. 2007-02. Guidance for Completing 2007 PARTs. Memorandum to OMB Program Associate Directors, OMB Program Deputy Associate Directors, Agency Budget and Performance Integration Leads, Agency Program Assessment Rating Tool Contacts, from Diana Espinosa, Deputy Assistant Director for Management, Office of Management and Budget, Executive Office of the President, Washington, DC. January 29, 2007. Attachment: Guide to the Program Assessment Rating Tool (PART). January 2007 [online]. Available: http://stinet.dtic.mil/ cgi-bin/GetTRDoc?AD=ADA471562&Location=U2&doc=GetTRDoc.pdf [accessed Nov. 7, 2007].

---

[7]Some examiners do have training and experience in science or engineering, but this is not a requirement for the position.

Appendix A

Biographic Information on the Committee on Evaluating the Efficiency of Research and Development Programs at the U.S. Environmental Protection Agency

Gilbert S. Omenn (Chair) is professor of internal medicine, human genetics, and public health and director of the university-wide Center for Computational Medicine and Biology at the University of Michigan. He served as executive vice president for medical affairs and as chief executive officer of the University of Michigan Health System from 1997 to 2002. He was dean of the School of Public Health and professor of medicine and environmental health at the University of Washington at Seattle from 1982 to 1997. His research interests include cancer proteomics, chemoprevention of cancer, public-health genetics, science-based risk analysis, and health policy. He was principal investigator of the Carotene and Retinol Efficacy Trial (CARET) of preventive agents against lung cancer and heart disease, director of the Center for Health Promotion in Older Adults, and creator of the university-wide initiative Public Health Genetics in Ethical, Legal, and Policy Context at the University of Washington and the Fred Hutchinson Cancer Research Center. He served as associate director of the Office of Science and Technology Policy and associate director of the Office of Management and Budget in the Executive Office of the President in the Carter administration. He is a long-time director of Amgen Inc. and of Rohm & Haas Company. He is a member of the Council and leader of the Plasma Proteome Project of the international Human Proteome Organization (HUPO). He was president of the American Association for the Advancement of Science during 2005-2006. Dr. Omenn is the author of 430 research papers and scientific reviews and author or editor of 18 books. He is a member of the Institute of Medicine, the American Academy of Arts and Sciences, the Association of American Physicians, and the American College of Physicians. He chaired the

Presidential/Congressional Commission on Risk Assessment and Risk Management (the "Omenn Commission"), served on the National Commission on the
Environment, and chaired the National Academies Committee on Science, Engineering, and Public Policy and the National Research Council Board on Environmental Studies and Toxicology.

George V. Alexeeff is deputy director for scientific affairs in the Office of Environmental Health Hazard Assessment (OEHHA) of the California Environmental Protection Agency. He oversees a staff of over 80 scientists in multidisciplinary evaluations of the health impacts of pollutants and toxicants in air,
water, soil, and other media. The office's activities include reviewing epidemiologic and toxicologic data to identify hazards and derive risk-based assessments, developing guidelines to identify chemicals hazardous to the public, rec-
ommending air-quality standards, identifying toxic air contaminants, developing
public-health goals for water contaminants, preparing evaluations for carcinogens and reproductive toxins, issuing sport-fishing advisories, training health
personnel on pesticide-poisoning recognition, reviewing hazardous-waste site
risk assessments, and conducting multimedia risk assessments. He was chief of
the Air Toxicology and Epidemiology Section of OEHHA from 1990 to 1998.
Dr. Alexeeff has over 50 publications in toxicology and risk assessment. He
recently served on the National Research Council Committee to Review the
OMB Risk Assessment Bulletin. Dr. Alexeeff earned his PhD in pharmacology
and toxicology from the University of California, Davis.

Radford Byerly Jr. is a research scientist at the Center for Science and Technology Policy Research, University of Colorado. He formerly worked at the
National Institute of Standards and Technology (then the National Bureau of
Standards) in environmental measurement and fire research, served as chief of
staff of the U.S. House of Representatives Committee on Science and Technology, and was director of the University of Colorado's Center for Space and Geo-
sciences Policy. He served as a member of the National Aeronautics and Space
Administration Space Science and Space Station Advisory Committees and
served on National Science Foundation site-visit committees and review panels.
Dr Byerly is a member of the National Research Council Space Studies Board
and served on the Committee on the Scientific Context for Space Exploration
(2004-2005), the Committee on Principles and Operational Strategies for Staged
Repository Systems (2001-2003), the Committee on Building a Long-Term Environmental Quality Research and Development Program in the U.S. Department of Energy (2000-2001), and the Board on Assessment of National Institute
of Standards and Technology Programs (1995-2000).

Edwin H. Clark II is a Senior Fellow at the Earth Policy Institute in Washington, DC. He is a former president of Clean Sites Inc. in Alexandria, VA, and
former secretary of natural resources and environmental control for the state of
Delaware. He was vice president of the Conservation Foundation and associate

assistant administrator for pesticides and toxic substances in the U.S. Environmental Protection Agency (EPA). He has served as a member of the National Research Council Board on Environmental Studies and Toxicology and on several committees, including the Committee to Evaluate the Science, Engineering, and Health Basis of the DOE's Environmental Management Program; the Committee on Risk-Based Criteria for Non-RCRA Hazardous Waste; the Committee to Review EPA's Research Grants Program; the Committee on Superfund Site Assessment and Remediation in the Coeur D'Alene River Basin; and the Committee to Review the Worker and Public Health Activities Program Administered by the Department of Energy and the Department of Health and Human Services. He holds a PhD in applied economics from Princeton University.

Susan E. Cozzens is a professor in the School of Public Policy at Georgia Institute of Technology. Until recently, she was chair of the School of Public Policy at the Georgia Institute of Technology; she left that position in 2003 to focus on her research activities. From 1995 through 1997, Dr. Cozzens was director of the Office of Policy Support at the National Science Foundation. She has served as a consultant to the National Academies Committee on Science, Engineering, and Public Policy, the Office of Science and Technology Policy, the National Science Foundation, the Institute of Medicine, the Office of Technology Assessment, the General Accounting Office, the National Cancer Institute, the National Institute on Aging, the National Institutes of Health, and the National Institute for Occupational Safety and Health and on advisory committees for the American Association for the Advancement of Science (AAAS) (Liberal Education and the Sciences and EPSCOR Evaluation), the National Academy of Sciences (NSF Decision-Making for Major Awards), and the Office of Technology Assessment (Human Genome Project). She has been an invited speaker on science policy and research evaluation at the Ministry for Research and Technology in France, the Research Council of Norway, the Institute for Policy and Management in Beijing, and the Fundamental Science Foundation of Sao Paulo, Brazil, and is incoming chair of the AAAS Committee on Science, Engineering, and Public Policy. Her PhD is in sociology from Columbia University (1985) and her bachelor's degree from Michigan State University (1972). She is a recipient of Rensselaer Polytechnic Institute's Early Career Award, a member of Phi Beta Kappa and Phi Kappa Phi, and a Fellow of AAAS.

Linda J. Fisher is vice president and chief sustainability officer for DuPont. She has responsibility for advancing DuPont's progress in achieving sustainable growth, DuPont environmental and health programs, the company's product-stewardship programs, and global regulatory affairs. She joined DuPont in 2004. Before joining DuPont, Ms. Fisher served in a number of key leadership positions in government and industry, including deputy administrator of the Environmental Protection Agency (EPA); EPA assistant administrator, Office of Prevention, Pesticides, and Toxic Substances; EPA assistant administrator, Office of Policy, Planning, and Evaluation; and chief of staff to the EPA administrator. Ms. Fisher, an attorney, was also vice president of government affairs for Monsanto and counsel with the Washington, DC, law firm Latham and Watkins. She is a member of the DuPont Health Advisory Board and the DuPont Biotechnology Advisory Panel and serves as liaison to the Environmental Policy Committee of the DuPont Board of Directors. Ms. Fisher serves on the Board of Directors of the Environmental Law Institute and on the Board of Trustees of the National Parks Foundation. She received a JD from Ohio State University and an MBA from George Washington University.

J. Paul Gilman is director of the Oak Ridge Center for Advanced Studies, Oak Ridge National Laboratory. Previously, he served as assistant administrator for research and development at the U.S. Environmental Protection Agency. He also worked at the Office of Management and Budget, where he had oversight responsibilities for the Department of Energy (DOE) and all other science agencies, and at DOE, where he advised the secretary of energy on scientific and technical matters. From 1993 to 1998, Dr. Gilman was the executive director of the Commission on Life Sciences of the National Research Council. He is a member of the National Research Council Board on Environmental Studies and Toxicology. Dr. Gilman earned PhDs in ecology and evolutionary biology from Johns Hopkins University.

T.J. Glauthier is head of TJG Energy Associates, where he provides consulting and executive advisory services to clients in the energy sector, including venture-capital companies, private-equity investors, alternative-energy companies, electric utilities, and global energy companies. He serves on the Board of Directors of Union Drilling, Inc., EnerNOC, Inc., and EPV Solar, Inc. He is an advisor to Foundation Capital LLC, a venture capital firm in Silicon Valley. He also advises the partners and clients in Booz Allen Hamilton's global energy sector management consulting practice. His pro bono activities include serving as an adviser to Stanford University's Precourt Institute for Energy Efficiency and on the Board of Directors of the San Mateo County Resource Conservation District. From 2001 to 2004, Mr. Glauthier was president and CEO of the Electricity Innovation Institute, an affiliate of EPRI. He was the deputy secretary and COO of the U.S. Department of Energy (DOE) from 1999 to 2001. For 5 years before going to DOE, he served in the White House as associate director for natural resources, energy, and science in the Office of Management and Budget. Earlier, Mr. Glauthier was a vice president of Temple, Barker & Sloane, a management consulting firm. Immediately before joining the Clinton administration, he spent 3 years as director of energy and climate change at the World Wildlife Fund, focusing on technology transfer, the climate-change treaty, and the 1992 Earth Summit in Rio de Janeiro. Mr. Glauthier is a graduate of Claremont McKenna College and the Harvard Business School.

Carol J. Henry is an independent consultant, having retired as vice president of industry performance programs at the American Chemistry Council (ACC). She has served as vice president for science and research at ACC, managing and guiding the Long-Range Research Initiative. Previously, Dr. Henry served as director of the Health and Environmental Sciences Department of the American Petroleum Institute, as associate deputy assistant secretary for science and risk policy at the U.S. Department of Energy, as director of the Office of Environmental Health Hazard Assessment (OEHHA) at the California Environmental Protection Agency, and as executive director of the International Life Sciences Institute's Risk Science Institute. A diplomate of the American Board of Toxicology, Dr. Henry is a member of the American College of Toxicology, of which she has been president; the Society of Toxicology; and the American Chemical Society, where she was elected to the Board of Managers of the Chemical Society of Washington. Dr. Henry was a member of the National Research Council Board on Environmental Studies and Toxicology and most recently a member of its Committee on Human Biomonitoring of Environmental Chemicals. She serves on the Roundtable on Environmental Health Sciences, Research, and Medicine of the Institute of Medicine; on the *Environmental Health Perspectives* Editorial Review Board; and as cochair of the Science Advisory Board for the Harvard School of Public Health-Cyprus International Initiative for the Environment and Public Health. Dr. Henry received her PhD in microbiology from the University of Pittsburgh.

Robert J. Huggett is a consultant and professor emeritus of marine science at the College of William and Mary. From 1997 to 2004, he served as professor of zoology and vice president for research and graduate studies at Michigan State University. Dr. Huggett's aquatic-biogeochemistry research involved the fate and effects of hazardous substances in aquatic systems. From 1994 to 1997, he was the assistant administrator for research and development for the U.S. Environmental Protection Agency, where his responsibilities included planning and directing the agency's research program. He has served on the National Research Council Board on Environmental Studies and Toxicology. Dr. Huggett earned his PhD at the College of William and Mary.

Sally Katzen is visiting professor of law at George Mason University School of Law. She has taught administrative law and information-technology policy at the University of Michigan Law School, administrative law at the University of Pennsylvania Law School and the Georgetown Law Center, and American government at Smith College, Johns Hopkins University, and the University of Michigan (Washington Program). Before her teaching positions, she served as the administrator of the Office of Information and Regulatory Affairs in the Office of Management and Budget (OMB) in 1993-1998, as the deputy director of the National Economic Council in the White House in 1998-1999, and as the deputy director for management in OMB in 1999-2001. Before her government service, she was a partner in the Washington, DC, law firm of Wilmer, Cutler, and Pickering, specializing in administrative law and legislative matters. Ms. Katzen recently served on the National Research Council Committee to Review the OMB Risk Assessment Bulletin. She earned her JD from the University of Michigan Law School.

Terry F. Young is an independent consultant, working primarily on behalf of nonprofit environmental organizations. Her recent work includes the development of a system that uses economic incentives, including input pricing and tradable discharge permits, to control farm pollution in California's San Joaquin Valley. Additional work includes development of ecologic indicators to track management and restoration of ecologic systems. Dr. Young has published on economic incentives for environmental protection, indicators of ecologic integrity, and market solutions for water pollution. She recently was appointed by Governor Schwarzenegger to the California Regional Water Quality Control Board, San Francisco Region. Dr. Young is a member of the Environmental Protection Agency (EPA) Science Advisory Board and served as a member of the National Research Council committee to review EPA's research-grants program. Dr. Young received her PhD in agricultural and environmental chemistry from the University of California, Berkeley.

Appendix B

Evaluating the Efficiency of Research and Development Programs at the Environmental Protection Agency: Workshop Summary

With oversight by the Committee on Science, Engineering, and Public Policy and the Board on Environmental Studies and Toxicology, the Committee on Evaluating the Efficiency of Research and Development Programs at the U.S. Environmental Protection Agency (EPA) organized a public workshop in April 2007 (the full presentations made at the workshop are available in the Public Access File of the National Research Council created for this project).

Representatives of the Office of Management and Budget (OMB), the EPA, other federal agencies that perform research, and industry addressed the following questions in the context of the Government Performance and Results Act of 1993 (GPRA) and the Program Assessment Rating Tool (PART):

1. What efficiency measures are currently used for EPA R&D programs and other federally funded R&D programs?

2. Are the efficiency measures sufficient? Are they outcome-based?

3. What principles should guide the development of efficiency measures for federally funded R&D programs in general?

4. What efficiency measures should be used specifically for EPA's basic and applied R&D programs?

PRESENTATION BY OFFICE OF MANAGEMENT AND BUDGET STAFF

The rationale for using PART is that taxpayers deserve to have their money spent wisely to create the maximal benefit. OMB developed PART in 2002 because the reporting process associated with GPRA was losing momentum. The office wanted another opportunity to focus on defining success and on the design and implementation of programs. OMB has decided that efficiency should be measured because R&D programs need to maintain a set of high-priority, multi-year objectives with annual performance outputs and milestones that show how one or more outcomes will be reached despite limited resources.

PART is used for several purposes. The most basic is to evaluate the success of programs. The second is to monitor the annual improvement plans required of each program.

Evaluation of the first two PART criteria, quality and relevance, primarily by expert review, has caused few problems. Application of the performance criterion—especially the measures of efficiency—has proved to be a challenge. As a result, OMB has approached implementation of that third criterion as a learning process.

The two relevant PART questions concerning the efficiency of R&D are questions 3.4 and 4.3. Question 3.4 asks, "Does the program have procedures (e.g., competitive sourcing/cost comparisons, IT improvements, appropriate incentives) to measure and achieve efficiencies in program execution?" A way to measure efficiency is required for a "yes" response. Question 4.3 asks, "Does the program demonstrate improved efficiencies or cost effectiveness in achieving program goals each year?" Answering question 4.3 is predicated on a "yes" response to question 3.4 and improved efficiencies or cost effectiveness in achieving goals should be described in terms of dollars when possible.

For the President's Management Agenda (PMA), the agency is given a "yellow" score when at least 50% of agency programs rated by PART have at least one efficiency measure and a "green" score when all agency programs rated by PART have at least one efficiency measure.

The meaning of efficiency, as OMB has applied PART, includes both outcomes or outputs for a given amount of inputs and inputs for a given amount of outcomes or outputs. Outcome efficiency might be measured in the economic terms of benefit-cost ratio, cost-benefit ratio, or cost effectiveness. Output efficiency might be measured in terms of productivity (input/output) or unit cost (output/input) or with respect to a standard or benchmark.

For outcomes, attribution of success or failure is inexact and may be based on indicators as diverse as improved targeting of beneficiaries or customers, a radically different mode of intervention, productivity improvements, or cost reductions. For outputs, efficiency might be described in relation to a program's resources, such as the use of labor or material, improved capability, or procurement. PART also requires that measures of outcome efficiency "consider the benefit to the customer and serve as an indicator of the program's operational performance."

Output efficiencies have various potential criteria. They must reflect efficient use of resources, measure changes over time that should correspond to a decrease or increase in related costs, and include an assessment of the comparability of the kinds of outputs produced.

OMB also attached high priority to assessing R&D programs at the project level; that is, there must be single-year and multiyear R&D objectives, with annual performance outputs, to track how a program (an aggregation of projects) will improve scientific understanding and its application. Programs must also provide schedules with annual milestones for future competition, decision, and termination points, highlighting changes from previous schedules. The problem is that basic R&D does not fit those criteria, and much applied R&D does so only with difficulty.

OMB has suggested the use of "earned-value management" (EVM) as a technique for tracking R&D efficiency. EVM plots expenditures against time, beginning with actual cost in dollars and comparing it with current earned value and planned value. EPA has agreed to use EVM as an efficiency-assessment tool on a pilot basis, although no agency is using it for basic research.

OMB sees several difficulties in applying PART to research. One is the concern that new PART requirements will cause agencies to favor research that fits those measures and to defund research that does not fit them. Furthermore, OMB has found it hard to devise efficiency measures for research that can be used to identify improvement each year, as is expected generally under PART. OMB's view is that although EVM is effective for parts of programs, such as construction projects, it is difficult to apply it to entire R&D programs.

PRESENTATION BY ENVIRONMENTAL
PROTECTION AGENCY STAFF

EPA representatives described the difficulties presented in finding an appropriate way to evaluate the performance of their research programs, especially with respect to the efficiency criterion. As a result, OMB gave EPA a "yellow" rating for the Budget and Performance Integration initiative under the PMA. After experimenting with several possibilities, the agency decided to use the number of peer-reviewed papers published per full-time equivalent (FTE) as an efficiency measure for its Water Quality Research Program (WQRP). The PART Appeals Board ruled that EPA could use publications on condition that the WQRP develop an "outcome-oriented efficiency measure." That agreement helped EPA achieve a "green" rating in March 2007 on the PMA.

EPA recognized the limitation of using publication citations, seen as better for measuring productivity than for measuring efficiency. One issue is the quality of the publications. Another is that publications are not a useful metric for many applied-research programs, especially in ecologic fields, in which researchers publish fewer papers than in, for example, toxicology. A major problem in applying any single metric across even a single agency is the variation among programs. A large percentage of the budget of the human-health program goes to extramural grants, which cannot be evaluated by the same measures as EPA's extensive inhouse laboratory system.

EPA did note that its Office of Research and Development (ORD) does a bibliographic analysis of every program. It also quantifies the extent to which ORD research is used to support regulations. Other evaluation tools are client surveys and the average time spent in producing assessments. EPA considered other measures, such as research vs overhead and citations per dollar invested. For ecologic research, it tried a "version of EVM," comparing projected costs for a long-term goal with actual costs in the context of scheduled output for the goal. A problem is that goals are planned on a multiyear cycle and are difficult to measure annually. OMB would not accept the use of expert review to measure efficiency.

A problem for every agency is that OMB examiners vary widely in their knowledge of research and their views of what is acceptable. One examiner may accept a particular efficiency measure for multiple programs that another does not accept.

Discussion focused on the concern that budget allocations might shift in the direction of "efficient" programs with little regard for the quality of the science. EPA acknowledged that low PART scores sometimes mean less money for a program. Particularly vulnerable was basic research or a long-term study with unclear outcomes, such as the search for a causal connection between drinking-water quality and cancer, in which the agency has been looking for "proxies" that have logical linkage to outcomes.

With regard to negotiating the application of EVM, it has been applied as a short-term solution. EPA's intention was to work out alternative measures that would work not just for EPA but also for other research-funding agencies.

In response to a question as to whether EPA could align the progress of a multiyear program with the budget, EPA noted that each long-term research plan is revised and updated annually by a research-coordination team. EPA relies on customer surveys and decision documents to indicate how results of research are used.

PRESENTATION BY DEPARTMENT OF ENERGY STAFF

Staff of the Department of Energy (DOE) described the impact of PART on the many activities of the DOE Office of Science. In 2002, DOE received low PART scores (50% and 60%) because of its performance measures. The agency revised its system and raised its scores to the 80s and 90s.

For evaluating the quality and relevance of R&D, DOE depends on peer review by committees of visitors. It had not reviewed performance before the creation of PART, so it established a committee to test appropriate metrics. It tried using the number of hours that large DOE facilities were available to users, but because the facilities were all fully subscribed, this metric was not useful for annual improvement. OMB asked for a new measurement involving dollars per unit of work; after long discussions, DOE responded that the use of a single unit-per-dollar measure would not be effective. Instead, it proposed a detailed examination of management procedures to be reviewed regularly by expert reviewers knowledgeable about the processes.

DOE uses EVM for the construction phase of a facility but does not apply it to R&D or the operation of facilities. DOE has reviewed the practices of other agencies and major corporations and found no useful models. No one knew how to define *value* for the kinds of projects in the DOE portfolio, so performance could not be established by using a dollar value. Therefore, DOE turned again to expert reviewers and asked them to quantify the value of a project, assign risk and probability curves, and then conduct EVM analysis.

DOE noted that the director of the Office of Science and Technology Policy had set up a committee on this issue. Its assignment is to examine the literature for ways to identify prospective benefits of research, "something no one is presently able to do," and to seek input from all federal agencies on useful tools for evaluating research. The charge is to describe a mechanism for measuring the value of research and to assign a cost to compliance. Neither GPRA nor PART addresses agencies' costs to comply with the data-gathering, analytic, and reporting requirements, which can be considerable. Other agencies also expressed concern about the time and budgetary costs of compliance, including a statement by the National Institutes of Health (NIH) staff that 250 people worked full-time for 3 months to comply with PART for the NIH extramural program.

For DOE, PART has a natural application in engineering and other predictable processes. Research represented a modest part of all the R&D work done, and the direction and outcomes were never as clear and specific as building a facility. One goal of OMB was to draw a boundary around administrative costs and reduce them. For example, one measure at DOE is to maintain total administrative overhead costs in relation to total program costs at 12%. But DOE recognized the trap of attempting to drive down administrative costs continuously.

As one example, the Energy Efficiency and Renewable Energy program uses an overhead metric but only for operational and construction programs, not for R&D. The program also experimented with using the relationship between the corporate program-management line and the total program R&D budget but found it to be "ungameable."

For R&D, DOE uses the "alternative efficiency measure" of peer review for all portfolios every 2nd or 3rd year. It is fairly cost-effective, allowing the agency to look at what is proposed and how well it is performed, identify ideas that lack merit, discontinue inefficient processes, redirect R&D, or terminate a poorly performing project.

PRESENTATION BY NATIONAL SCIENCE FOUNDATION STAFF

The National Science Foundation (NSF) evaluates its programs, using strategic outcome goals (discovery, learning, and research infrastructure) and annual assessments of its investments in long-term research. That is done by an external expert Advisory Committee for GPRA Performance Assessment. It reviews program accomplishments foundation-wide and submits a report to the director with conclusions and recommendations.

NSF has also initiated a new annual stewardship-goals assessment with eight annual performance goals. The assessment focuses on proposal processes, program administration, and management.

A well-known NSF approximation of a measurement consists of the program-portfolio level assessments performed every 3 years by external committees of visitors. This process, called merit review, is a detailed and long examination of both technical merit and broader impacts of research.

NSF tracks efficiency primarily in two ways. One is to measure the time to decision on research awards, which is important to researchers who depend on grant support. NSF is able to inform 70% of applicants within 6 months. The second is to measure facility cost, schedule, and operation. A goal for new facilities is to keep cost overruns and schedule variances for construction to less than 10% of the approved project plan for 90% of the facilities, and a parallel goal for operating facilities is to keep operating time lost because of unscheduled downtime to less than 10% for 90% of the facilities.

PRESENTATION BY NATIONAL INSTITUTES OF HEALTH STAFF

NIH created an Office of Portfolio Analysis and Strategic Initiatives to examine systemic assessments and practice the "science of science management." Two general points emerged: the difficulty of using a business-model approach to measure efficiency in science and NIH success in using PART on research and research-support activities. Some 99% of the NIH portfolio had been "PART-ed"; 95% of programs were rated effective, and the other 5% were rated moderately effective. The review of extramural research is limited to elements of the program under NIH's management control.

With respect to both the extramural and intramural programs, NIH claims some success in improved management. The extramural-research program has achieved cost savings through improved grant administration. The intramural-research program has saved money by reallocating laboratory resources. The building and facilities program has monitored its property condition index. The extramural construction program has saved funds by expanding the use of electronic management tools.

In the business model approach used by PART, efficiency has three aspects: time, cost, and deliverables. Efficiency can be increased by improving any one of them as long as the other two do not worsen. In scientific discovery, however, variables are largely unknown; because the outcome is unpredictable knowledge, the inputs of time, cost, and resources are difficult to estimate. Some inputs may also be fixed by scientific methods. If the goal of a project is to produce a microarray, deliverables cannot be "increased" by producing two or three microarrays.

There are other reasons why science does not fit easily with this type of business model. In business, risk is usually undesirable; in research, whether in the private sector or the public sector, high-risk projects are strongly associated with breakthrough innovative outcomes. Nor does the business model capture the null hypothesis, which states that a negative result gives valuable information. Changing direction in a project may look like poor management, but it may be good science. The outcome may be unexpected or lead to an unexpected benefit, as occurs with drug benefits. If multiple teams are doing the same research, there is no way to calculate the relative value of each team. Finally, because of the government's public-health responsibilities, including such issues as rare diseases, costs and benefits are different from those in for-profit enterprises whose measure is new product sales.

PRESENTATION BY NATIONAL AERONAUTICS
AND SPACE ADMINISTRATION STAFF

The National Aeronautics and Space Administration (NASA) has focused on aligning the PART process and the annual GPRA process to yield a single set of externally reported measures. That had allowed it, for example, to link the monitoring of mission cost and schedule performance metrics and GPRA outcomes.

Recently, NASA has moved away from agencywide measures of efficiency toward program-specific measures. The future focus, in complying with PART, is on finding efficiencies in operational activities and supporting business processes that lead to science and R&D products. NASA is using PART measures in the complex launch process, for example, and to find safe ways to reduce the size of the Space Shuttle workforce. It plans to use them in other ways, such as increasing the on-time availability and operation of ground test facilities and reducing the cost per minute of network support for space missions.

NASA has found that the definitions and guidance for PART efficiency measures are most useful for repetitive, stable, and baselined processes and for some aspects of the management of R&D, such as financial management, contracting, travel-processing, and capital-assets tracking. But for long-term research, NASA is unable, for instance, to put an efficiency measure on finding the dark matter of the universe. Much of what NASA does is make discoveries and prototypes on unrepeatable time scales dictated by science. NASA's efficiency measures tend to be process-oriented, not outcome-oriented. NASA urged more flexibility for the process—for example, to recognize that short-term decreases in efficiency might lead to long-term efficiency gains and to recognize the need to balance effectiveness and efficiency.

PRESENTATION BY NATIONAL INSTITUTE FOR
OCCUPATIONAL SAFETY AND HEALTH STAFF

The National Institute for Occupational Safety and Health (NIOSH) is not a regulatory body; it serves as the research partner of the Occupational Safety and Health Administration in the Department of Labor, although it is organizationally part of the Centers for Disease Control and Prevention in the Department of Health and Human Services. It uses independent expert review to evaluate its 30 research programs, which exist in a "matrix" with substantial overlaps. The programs are relatively small, with budgets of $5-35 million. Eight NIOSH programs are being studied by other ad hoc committees of the National Research Council for relevance, impact, and emerging issues.

NIOSH research results can be divided into outputs, intermediate outcomes, and outcomes:

- Outputs include peer-reviewed publications, NIOSH publications, communications to regulatory agencies or Congress, research methods, control technologies and patents, and training and information products.
- Intermediate outcomes include regulations, guidance, standards, training and education programs, and pilot technologies.
- End outcomes include reductions in fatalities, injuries, illnesses, and exposures to hazards.

Some efficiency measures are used for PART, beginning with percentage of grant award and funding decisions made available to applicants within 9 months while a credible and efficient peer-review system is maintained.

NIOSH has considered its own principles for progress on research-program efficiency measures, including the degree of control over efficiency variables, refinements of all PART definitions for R&D, and the "need for impacts" to drive efficiency.

Several potential efficiency measures have emerged, including

- Correlation between research-activity funding and congruence of activity goals over time.
- Correlation between funding and number, quality, representativeness, and potential value of research partnerships over time.
- The correlations above with the use of research.

PRESENTATION BY PROCTER & GAMBLE STAFF

Procter & Gamble (P&G) maintains a considerable middle-term and long-term research effort in hazard characterization, risk assessment, and development of core competences. Efficiency as measured by time to market is critical for firms such as P&G. The impact on corporate profits of being the first-to-market can be substantial. P&G's short-term research supports new product initiatives and investigates unusual toxicity. Most of its efficiency measures are designed to save time in product development, increase confidence about safety, and build external relations (although this is not quantifiable).

PRESENTATION BY IBM STAFF

The company uses efficiency measures for some kinds of activities, such as

- Return on investment in the summer internship program and graduate fellowship program: What percentage of the recipients return as regular IBM research employees?
- A "Bureaucracy Busters" initiative to reduce bureaucracy in laboratory support, information-technology support, human-resources processes, and business processes.
- Tracking of the patent-evaluation process.
- Customer-satisfaction surveys to evaluate the effects of service reductions.
- Measurement of response time and turnaround time for external contracts.
- Measurement of span-of-responsibility for secretarial support.

IBM representatives agreed that basic research is hard to measure and that the structure of EVM was almost antithetical to the performance of basic research. By EVM standards, surprise is bad. In basic research, surprise is good. EVM is oriented toward projects, not exploratory work in which an answer is not self-evident at the beginning.

Some intrinsic challenges in assessing basic research are to define value and to specify its recipients. It is desirable to measure outcomes rather than outputs because outcomes are a "cleaner" effectiveness measure and have a "clear value." In measuring research on water quality, however, outcomes are unknowable, and such an output as the number of publications per FTE may be the best approach available. Evaluating the quality of research is not hard, but evaluating the impact of research is much more difficult for a corporation until a place is established in the market.

PRESENTATION BY DOW CHEMICAL CORPORATION STAFF

Dow spends only a small percentage of its R&D budget on science, and it is aimed primarily at ensuring that products will not harm human health or the environment. Inhouse expertise is supported for several reasons:

- To maintain state-of-the-art competence.
- To help to translate innovations into use by business customers.
- To benchmark to external standards of cost, timing, and quality.

In research, Dow works to exploit laboratory-integrated research strengths, including toxicity testing, analytic research, and mode-of-action research. The company also collaborates with various research partners to gain access to new technology and expertise and to enhance credibility through publication and participation in the scientific community.

PRESENTATION BY ALCOA STAFF

Alcoa spends about 1% of sales on research (of which 75% is inhouse) and is just now beginning to look at efficiency measures. For example, a return-on-investment calculation would include the following:

- Variable cost improvement.
- Margin impact from organic growth.
- Capital avoidance.
- Cost avoidance.

The annual impact of those four metrics over a 5-year period is compared with the total R&D budget. The resulting metric is used to evaluate the overall value of R&D programs and the current budget focus. Although the reported numbers constitute a lagging indicator, the company tries to encourage business case development and projects expected financial impact on current and future projects whenever possible. Projects with a clear path to value creation are more likely to be funded than projects with no clear business gains.

Possible measures to improve efficiency presented by Alcoa include the following:

- For greater up-front business-case development:
  - Identify the customer.
  - Apply customer support and commitment.
  - Use a rigorous process for value capture.
  - Insist on transparency.
- For a stage-gate process:
  - Establish objectives and timetables.
  - Require completion before additional funding.
- For project review:
  - Have periodic review by a mix of supporters and skeptics to test objectives, feasibility, progress, and potential for success.

- For program review:
  - Aggregate R&D expenditures by laboratory group or identifiable programs and publish value capture or success rate for each annually.

Appendix C

Program Assessment Rating Tool (PART) Questions[1]

OFFICE OF MANAGEMENT AND BUDGET

Section I. Program Purpose and Design

1.1: Is the program purpose clear?

1.2: Does the program address a specific and existing problem, interest, or need?

1.3: Is the program designed so that it is not redundant or duplicative of any other Federal, State, local or private effort?

1.4: Is the program design free of major flaws that would limit the program's effectiveness or efficiency?

1.5: Is the program design effectively targeted so that resources will address the program's purpose directly and will reach intended beneficiaries?

Section II. Strategic Planning

2.1: Does the program have a limited number of specific long-term performance measures that focus on outcomes and meaningfully reflect the purpose of the program?

2.2: Does the program have ambitious targets and timeframes for its long-term measures?

---

[1]OMB (Office of Management and Budget). 2007. Program Assessment Rating Tool Guidance No. 2007-02. Washington, DC.

2.3: Does the program have a limited number of specific annual performance measures that can demonstrate progress toward achieving the program's long-term goals?

2.4: Does the program have baselines and ambitious targets for its annual measures?

2.5: Do all partners (including grantees, sub-grantees, contractors, cost-sharing partners, and other government partners) commit to and work toward the annual and/or long-term goals of the program?

2.6: Are independent evaluations of sufficient scope and quality conducted on a regular basis or as needed to support program improvements and evaluate effectiveness and relevance to the problem, interest, or need?

2.7: Are budget requests explicitly tied to accomplishment of the annual and long-term performance goals, and are the resource needs presented in a complete and transparent manner in the program's budget?

2.8: Has the program taken meaningful steps to correct its strategic planning deficiencies?

Specific Strategic Planning Questions by Program Type

2.RG1: Are all regulations issued by the program/agency necessary to meet the stated goals of the program, and do all regulations clearly indicate how the rules contribute to achievement of the goals? (Regulatory)

2.CA1: Has the agency/program conducted a recent, meaningful, credible analysis of alternatives that includes trade-offs between cost, schedule, risk, and performance goals, and used the results to guide the resulting activity? (Capital Assets and Service Acquisition)

2.RD1: If applicable, does the program assess and compare the potential benefits of efforts within the program and (if relevant) to other efforts in other programs that have similar goals? (R&D)

2.RD2: Does the program use a prioritization process to guide budget requests and funding decisions? (R&D)

Section III. Program Management

3.1: Does the agency regularly collect timely and credible performance information, including information from key program partners, and use it to manage the program and improve performance?

3.2: Are Federal managers and program partners (including grantees, sub-grantees, contractors, cost-sharing partners, and other government partners) held accountable for cost, schedule and performance results?

3.3: Are funds (Federal and partners') obligated in a timely manner, spent for the intended purpose, and accurately reported?

3.4: Does the program have procedures (e.g., competitive sourcing/cost comparisons, IT improvements, appropriate incentives) to measure and achieve efficiencies and cost effectiveness in program execution?

3.5: Does the program collaborate and coordinate effectively with related programs?

3.6: Does the program use strong financial management practices?

3.7: Has the program taken meaningful steps to address its management deficiencies?

Specific Program Management Questions by Program Type

3.CO1: Are grants awarded based on a clear competitive process that includes a qualified assessment of merit? (Competitive Grants)

3.CO2: Does the program have oversight practices that provide sufficient knowledge of grantee activities? (Competitive Grants)

3.CO3: Does the program collect grantee performance data on an annual basis and make it available to the public in a transparent and meaningful manner? (Competitive Grants)

3.BF1: Does the program have oversight practices that provide sufficient knowledge of grantee activities? (Block/Formula Grant)

3.BF2: Does the program collect grantee performance data on an annual basis and make it available to the public in a transparent and meaningful manner? (Block/Formula Grant)

3.RG1: Did the program seek and take into account the views of all affected parties (e.g., consumers; large and small businesses; State, local and tribal governments; beneficiaries; and the general public) when developing significant regulations? (Regulatory)

3.RG2: Did the program prepare adequate regulatory impact analyses if required by Executive Order 12866, regulatory flexibility analyses if required by the Regulatory Flexibility Act and SBREFA, and cost-benefit analyses if required under the Unfunded Mandates Reform Act; and did those analyses comply with OMB guidelines? (Regulatory)

3.RG3: Does the program systematically review its current regulations to ensure consistency among all regulations in accomplishing program goals? (Regulatory)

3.RG4: Are the regulations designed to achieve program goals, to the extent practicable, by maximizing the net benefits of its regulatory activity? (Regulatory)

3.CA1: Is the program managed by maintaining clearly defined deliverables, capability/performance characteristics, and appropriate, credible cost and schedule goals? (Capital Assets and Service Acquisition)

3.CR1: Is the program managed on an ongoing basis to assure credit quality remains sound, collections and disbursements are timely, and reporting requirements are fulfilled? (Credit)

3.CR2: Do the program's credit models adequately provide reliable, consistent, accurate and transparent estimates of costs and the risk to the Government? (Credit)

3.RD1: For R&D programs other than competitive grants programs, does the program allocate funds and use management processes that maintain program quality? (R&D)

Section IV. Program Results/Accountability

4.1: Has the program demonstrated adequate progress in achieving its long-term performance goals?

4.2: Does the program (including program partners) achieve its annual performance goals?

4.3: Does the program demonstrate improved efficiencies or cost effectiveness in achieving program goals each year?

4.4: Does the performance of this program compare favorably to other programs, including government, private, etc., with similar purpose and goals?

4.5: Do independent evaluations of sufficient scope and quality indicate that the program is effective and achieving results?

Specific Results Questions by Program Type

4.RG1: Were programmatic goals (and benefits) achieved at the least incremental societal cost and did the program maximize net benefits? (Regulatory)

4.CA1: Were program goals achieved within budgeted costs and established schedules? (Capital Assets and Service Acquisition)

Appendix D

The Environmental Protection Agency's Strategic and Multi-year Planning Process

The Environmental Protection Agency (EPA) is guided by the principles of the President's Management Agenda to be "citizen-centered, results-oriented, and market-based" (EPA 2006, p. 149).

The EPA strategic plan delineates goals and describes how to achieve them, taking into account planning, budgeting, accountability, and performance measurements. Annual performance goals and measures are stated to track progress and achievements toward a long-term strategic goal. EPA's annual *Performance and Accountability Report* then assesses performance toward a particular goal that helps to delineate priorities and develop future budgets. Through evaluating performance measures to develop planning and decision-making, new environmental indicators are developed and described in the *Report on the Environment* (published every 4 years). The *Report on the Environment* further improves long-term measures in the strategic plan.

EPA's goals, measures, and accountability are advanced through accurate, timely environmental data. In the *Report on the Environment—Technical Document*, EPA provides a snapshot of current environmental conditions and a baseline against which accomplishments are measured. The environmental indicator, as described in the *Report on the Environment*, has facilitated identification of strategic goals, objectives and subobjectives, and strategic targets, which may be translated into measurable environmental results. Environmental trends show key concerns (data needs) and emerging issues and demonstrate the effectiveness of agencies in improving environmental quality. Budget and performance information, which is integrated in the strategic plan, can be used in long-term measures that meet Office of Management and Budget (OMB) Program Assessment Rating Tool (PART) requirements (EPA 2006, p. 150):

EPA collects and analyzes performance information against these measures to assess program performance over time and to evaluate the effectiveness of approaches to environmental problems. Based on these evaluations, we can adjust or modify our strategies to achieve better results. To encourage EPA staff and our partners to be accountable for delivering environmental results effectively and cost efficiently, we are also incorporating performance measures in EPA managers' performance agreements and, as appropriate, in our contracts, grants, and memoranda of understanding.

The strategic plan is developed in consultation with stakeholders and partners. In the development process, EPA managers organize meetings, participate in conferences and present briefings to assist with understanding, and receive commentary from stakeholders and partners. Through a long public-comment period and multiple revisions, an EPA strategic plan is adopted. Steps in the 2006-2011 strategic-plan development process included

- Early consultation on state and tribal issues and priorities.
- A draft architecture and full-draft release.
- Consultation with Congress and state and tribal partners.

Work is under way to update the strategic plan for 2009-2014. Since 2003, EPA has advocated five strategic goals. The current EPA strategic plan is for FY 2006-2011 and lists the following as goals and objectives:

- Goal 1: Clean Air and Global Climate Change
    - Objective 1.1: Healthier Outdoor Air
    - Objective 1.2: Healthier Indoor Air
    - Objective 1.3: Protect the Ozone Layer
    - Objective 1.4: Radiation
    - Objective 1.5: Reduce Greenhouse Gas Emissions
    - Objective 1.6: Enhance Science and Research
- Goal 2: Clean and Safe Water
    - Objective 2.1: Protect Human Health
    - Objective 2.2: Protect Water Quality
    - Objective 2.3: Enhance Science and Research
- Goal 3: Land Preservation and Restoration
    - Objective 3.1: Preserve Land
    - Objective 3.2: Restore Land
    - Objective 3.3: Enhance Science and Research
- Goal 4: Healthy Communities and Ecosystems
    - Objective 4.1: Chemical, Organism, and Pesticide Risks
    - Objective 4.2: Communities
    - Objective 4.3: Restore and Protect Critical Ecosystems

        –   Objective 4.4: Enhance Science and Research
- Goal 5: Compliance and Environmental Stewardship
  - Objective 5.1: Achieve Environmental Protection Through Improved Compliance
  - Objective 5.2: Improve Environmental Performance Through Pollution Prevention and Other Stewardship Practices
  - Objective 5.3: Improve Human Health and the Environment in Indian Country
  - Objective 5.4: Enhance Society's Capacity for Sustainability Through Science and Research (EPA 2006)

Within each goal, "emerging issues and external factors" are discussed; they identify probable challenges in the forthcoming years. In an effort to provide transparency to the public and other agencies, chapters of the strategic plan address the development of long-term measures and their relationship to annual performance measures used by OMB's PART. The strategic plan also discusses the development of indicators for EPA's *Report on the Environment.*

As an example of how the strategic plan is related to performance measures through the multi-year plan (MYP) and research plans, an objective of the goal 1 (Clean Air and Global Climate Change) is discussed below. For every objective in the strategic plan and MYP, there are subobjectives and targets. EPA uses MYPs for the research that works toward achieving the objectives in the strategic plan. It is usually at the MYP level that performance measures are evaluated. Research directions are used to develop research strategies, which are ultimately translated into performance measures in MYPs (EPA 2003a). An MYP is designed as a more detailed description of research and also embodies the goals of the ORD and EPA strategic plans (EPA 2003b).

> "Goal 1—Clean Air and Global Climate Change
>   'EPA is dedicated to improving the quality of the air Americans breathe' (EPA 2006, p. 11). To achieve this EPA develops regulations and programs with tribes, business, industry and other governments to reduce air pollution.
>   'Objective 1.1: Healthier Outdoor Air [listed as objective 1.1 Outdoor Air in MYP]' (EPA 2003b, p. 5) 'Subobjectives: Ozone and $PM_{2.5}$, Chronically Acidic Water Bodies and Air Toxics'" (EPA 2006, pp. 12-13).

For demonstration purposes, the "air toxics" subobjective is further detailed in the strategic-plan targets in the MYP and finally in the research plan and strategy. The strategic-plan goal for air toxics is

"By 2011, reduce the risk to public health and the environment from toxic air pollutants by working with partners to reduce air toxics emissions and implement area-specific approaches" (EPA 2006, p. 13).

To further refine the strategic-plan goal, a number of targets are listed. The targets for air toxics include

"By 2010, reduce toxicity-weighted (for cancer risk) emissions of air toxics to a cumulative reduction of 19 percent from the 1993 non-weighted baseline of 7.24 million tons."

"By 2010, reduce toxicity-weighted (for non-cancer risk) emissions of air toxics to a cumulative reduction of 55 percent from the 1993 non-weighted baseline of 7.24 million tons" (EPA 2006, p. 13).

To reach the strategic-plan targets, allow for performance measurement, and determine whether EPA is conducting valuable or appropriate research to reach the targets, the MYP is used. EPA established the following objective in the MYP for air toxics: "Through 2010, and consistent with established schedules, emissions of outdoor air pollutants will continue to decline, and ambient air quality will improve to or be maintained at levels that protect public health and the environment. Healthy air for the other pollutants will be maintained for the 123.7 million people that had healthy air in 2001" (EPA 2003b, p. 5). Again, to track progress in reaching the goal, the MYP establishes subobjectives. For air toxics, they are

"Control stationary sources of air toxics by using market-based and other regulatory programs to reduce emissions using the following target:"

"By 2007, federal air toxics regulations will reduce air toxics emissions by 2.2 million tons from their 1993 level of 3.7 million tons making absolute emissions reductions in air toxics compared to 2000 levels."

"Control mobile sources through federal regulations that will reduce air toxics emissions by 1.1 million tons from the 1996 level of 2.7 million tons."

"Reducing health risks and environmental effects from area source air toxics pollution found in localities including Indian country."

"Reduce air toxics risk at the local level by building on federally regulated emissions reductions."

"Milestones: 1) 2004 public release of the revised National Air Toxics Assessment based on the 1999 inventory and 2) Development of an air toxics monitoring program, and the ability to characterize and assess trends for 20% of the Indian tribes in 2010" (EPA 2003b, pp. 5-6).

For the subobjectives to be reached, EPA must undertake research that will define health risks associated with, environmental effects of, and methods of control of air toxics from different sources. To accomplish that task, EPA develops research strategies or plans with strategic principles that aid in guiding decisions and supporting ORD's research priorities as listed in the budget and MYP. For air toxics, a list of questions is used to identify which air toxics to study, what specific research to undertake, and the priorities in that research.

The research plan questions for the air-toxics subobjective are as follows (EPA 2002):

1. What are the sources of air toxics, and what are their characteristics?
2. What is the role of atmospheric transport, transformation, fate, and chemistry in air toxics concentrations (including indoor, micro-scale, urban, terrestrial, and regional concentrations)?
3. What is the relationship of concentrations of air toxics (from outdoor and indoor sources) to personal exposure?
4. What are the health hazards and dose-response relationships associated with exposure to air toxics?
5. What improvements can be made to dose-response assessments?
6. What health risks can be characterized quantitatively for people exposed to air toxics?
7. What risks from air toxics can be prevented and managed cost effectively?

A general subset of research activities arising from those questions consists of developing measurements, databases, methods, and biomarkers; validating models; identifying chemical mechanisms; and evaluating exposure concentrations.

REFERENCES

EPA (U.S. Environmental Protection Agency). 2002. Air Toxics Research Strategy. EPA-600-R-00-056. Office of Research and Development, U.S. Environmental Protection Agency, Washington, DC [online]. Available: http://www.epa.gov/ord/htm/documents/Air_Toxics.pdf [accessed Nov. 14, 2007].

EPA (U.S. Environmental Protection Agency). 2003a. 2003-2008 Strategic Plan: Direction for the Future. EPA-190-R-03-003. Office of Chief Planning Officer, Office of Planning Analysis and Accountability, U.S. Environmental Protection Agency, Washington, DC [online]. Available: http://www.epa.gov/epainnov/pdf/innovplan.pdf [accessed Nov 14, 2007].

EPA (U.S. Environmental Protection Agency). 2003b. Air Toxics Multi-Year Plan. Office of Research and Development, U.S. Environmental Protection Agency, Washington, DC [online]. Available: http://www.epa.gov/osp/myp/airtox.pdf [accessed Nov. 14, 2007].

EPA (U.S. Environmental Protection Agency). 2006. 2006-2011 Strategic Plan: Charting Our Course. EPA-190-R-06-001. Office of Chief Planning Officer, Office of Planning Analysis and Accountability, U.S. Environmental Protection Agency, Washington, DC [online]. Available: http://www.epa.gov/cfo/plan/2006/entire_report.pdf [accessed Nov. 14, 2007].

Appendix E

Agency and Industry Efficiency Measures

The table in this appendix (Table E-1) includes efficiency measures developed by agencies and industry. Most of the measures for the agencies were excerpted from the Office of Management and Budget (OMB) Program Assessment Rating Tool (PART) Web site (OMB 2007) and are related to programs of the "research and development" type. These measures either appear in the "Program Performance Measures" section or are cited as agency responses to questions 3.4 or 4.3 on the PART Web site. Question 3.4 is "Does the program have procedures (for example, competitive sourcing/cost comparisons, IT improvements, appropriate incentives) to measure and achieve efficiencies and cost effectiveness in program execution?" Question 4.3 is "Does the program demonstrate improved efficiencies or cost effectiveness in achieving program goals each year?" Some of the agency measures listed have been approved for use by OMB as of July 26, 2007.

The table is not an exhaustive list of efficiency measures used by the federal government, but it includes efficiency measures from a variety of agencies, including the Environmental Protection Agency (EPA), the Department of Energy (DOE), the U.S. Department of Agriculture (USDA), the Department of Commerce (DOC), the Department of Defense (DOD), the Department of Health and Human Services (DDHHS), the Department of the Interior (DOI), the Department of Labor (DOL), the Department of Transportation (DOT), the National Science Foundation (NSF), and the National Aeronautics and Space Administration (NASA). The industry efficiency measures (and a few agency efficiency measures) were gleaned from presentations during the April 2007 committee meeting.

TABLE E-1 Agency and Industry Efficiency Measures

Agency or Organization	Program	Year	Efficiency Measure
EPA	Endocrine Disruptors (combined EPA PART)	2004	(OPPTS) Cost per labor hour of contracted validation studies (EPA, unpublished material, April 23, 2007)
EPA	EPA Human Health Research	2005	Average time (in days) to process research-grant proposals from RFA closure to submittal to EPA's Grants Administration Division while maintaining a credible and efficient competitive merit-review system (as evaluated by external expert review) (EPA, unpublished material, April 23, 2007)
EPA	Land Protection and Restoration Research	2006	Average time (in days) for technical support centers to process and respond to requests for technical document review, statistical analysis, and evaluation of characterization and treatability study plans (EPA, unpublished material, April 23, 2007)
EPA	Water Quality Research	2006	Number of peer reviewed publications per FTE (EPA, unpublished material, April 23, 2007)
EPA	Human Health Risk Assessment Program	2006	Average cost to produce Air Quality Criteria/Science Assessment documents (EPA, unpublished material, April 23, 2007)
EPA	EPA Ecological Research	2007	Percentage variance from planned cost and schedule (approved 3/13/07) (EPA, unpublished material, April 23, 2007)
EPA	Drinking Water Research	2007	Percentage variance from planned cost and schedule (approved 3/13/07) (EPA, unpublished material, April 23, 2007)
EPA	PM Research	2007	Percentage variance from planned cost and schedule (approved 3/13/07) (EPA, unpublished material, April 23, 2007)
EPA	Global Change Research	2007	Percentage variance from planned cost and schedule (approved 3/13/07) (EPA, unpublished material, April 23, 2007)
EPA	Pollution Prevention Research	2007	Percentage variance from planned cost and schedule (approved 3/13/07) (EPA, unpublished material, April 23, 2007)

DOD	Defense Basic Research	2002	Long-term measure: portion of funded research chosen on basis of merit review; reduce non-merit-reviewed and determined projects by half in 2 years (from 6.0% to 3.0%) (OMB 2007)
DOE	Advanced Simulation and Computing	2002	Annual average cost per teraflops of delivering, operating, and managing all Stockpile Stewardship Program (SSP) production systems in given fiscal year (OMB 2007)
DOE	Coal Energy Technology	2005	Administrative costs as percentage of total program costs (OMB 2007)
DOE	Advanced Fuel Cycle Initiative	2003	Program direction as percentage of total R&D program funding (OMB 2007)
DOE	Generation IV Nuclear Energy Systems Initiative	2003	Program direction as percentage of total R&D program funding (OMB 2007)
DOE	National Nuclear Security Administration: Nonprolifera-tion and Verification Research and Development	2005	Cumulative percentage of active research projects for which independent R&D peer assessment of project's scientific quality and mission relevance has been completed during second year of effort (and again in each later 3-year period for projects found to be of merit) (OMB 2007)
DOE	Nuclear Power 2010	2003	Program direction as percentage of total R&D program funding (OMB 2007)
DOE	Basic Energy Sciences/ Biological and Environmental Research	2006	Average achieved operation time of scientific user facilities as percentage of total scheduled annual operation time; cost-weighted mean percentage variance from established cost and schedule baselines for major construction, upgrade, or equipment procurement projects (cost variance listed first) (OMB 2007)
DOE	Hydrogen Program	2003	In 2003, EERE Hydrogen Program had about 130 fuel-cell and hydrogen production research projects that were subject to *in-progress* peer review by independent experts

(Continued)

TABLE E-1 Continued

Agency or Organization	Program	Year	Efficiency Measure
			For all reviewed projects, reviewers provided written comments and numerical ratings — on a scale of 1-4, with 4 being highest — with resulting scores ranging of 2.2-3.9
			Program used review results to make important decisions to continue or discontinue projects
			Research efficiency = 1- [(no. of projects discontinued/(total no. of projects reviewed - no. of projects judged as completed - earmark projects)] (Beschen 2007)
DOI	U.S. Geological Survey – Biological Information Management and Delivery	2005	Average cost per gigabyte of data available through servers under program control (EPA, unpublished material, 2006)
DOI	U.S. Geological Survey – Biological Research & Monitoring	2005	Average cost per sample for selected high-priority environmentally available chemical analyses (EPA, unpublished material, 2006)
DOI	U.S. Geological Survey – Energy Resource Assessments	2007	Average cost of systematic analysis or investigation (dollars in millions) (EPA, unpublished material, 2006)
DOI	U.S. Geological Survey – Mineral Resource Assessment	2003	Average cost of systematic analysis or investigation; average cost per analysis allows comparisons among projects to determine how efficiencies can be achieved (EPA, unpublished material, 2006)
DOI	U.S. Geological Survey – Water Resources Research	2004	Average cost per analytic result, adjusted for inflation, is stable or declining over 5-year period (EPA, unpublished material, 2006)
DOI	U.S. Geological Survey – Water Information Collection and Dissemination	2004	Percentage of daily streamflow measurement sites with data that are converted from provisional to final status within 4 months of day of collection (EPA, unpublished material, 2006)

DOI	U.S. Geological Survey – Biological Research & Monitoring	2005	Percentage improvement in detectability limits for selected high-priority environmentally available chemical analytes (EPA, unpublished material, 2006)
DOI	U.S. Geological Survey – Geographic Research, Investigations, and Remote Sensing	2003	Percentage of total cost saved through partnering for data collection of high-resolution imagery (EPA, unpublished material, 2006)
DOI	Bureau of Reclamation – Science and Technology Program	2003	Each year, increase in R&D cost-sharing per reclamation R&D program dollar will contribute toward achieving long-term goal of 34% cumulative increase over 6-year period (OMB 2007)
DOT	Highway Research and Development/ Intelligent Transportation Systems	2004	Annual percentage of all research projects completed within budget (OMB 2007)
DOT	Highway Research and Development/ Intelligent Transportation Systems	2004	Annual percentage of research-project deliverables completed on time (OMB 2007)
DOT	Railroad Research and Development	2004	Organizational Excellence: Percentage of projects completed on time (OMB 2007)
Department of Education	National Assessment for Educational Progress	2003	Timeliness of NAEP data for Reading and Mathematics Assessment in support of President's No Child Left Behind initiative (time from end of data collection to initial public release of results for reading and mathematics assessments) (EPA, unpublished material, 2006)
Department of Education	National Center for Education Statistics	2003	NCES will release information from surveys within specified times; NCES collected baseline information in 2005, examining time-to-release for 31 recent surveys (National Assessment of Educational Progress releases not included in these figures) (EPA, unpublished material, 2006)

(Continued)

TABLE E-1 Continued

Agency or Organization	Program	Year	Efficiency Measure
DHHS	National Center for Health Statistics	2005	Number of months for release of data as measured by time from end of data collection to data release on Internet (OMB 2007)
DHHS	NIH Extramural Research Programs		By 2013, provide greater functionality and more streamlined processes in grant administration by continuing to develop NIH Electronic Research Administration System (eRA) (FY 2004) Develop plan to integrate OPDIVs into eRA (FY 2005) Integrate DHHS 50% of eligible DHHS OPDIVs as eRA users for administration of research grants (FY 2006) Integrate DHHS 100% of eligible DHHS OPDIVs as eRA users for administration of research grants Conversion of business processes – (FY 2005) 25% of business processes done electronically – (FY 2006) 40% – (FY 2007) 55% – (FY 2008) 80% (Duran 2007)
DHHS	NIH Intramural Research Program	2005	Reallocation of laboratory resources based on extramural reviews by Boards of Scientific Counselors (OMB 2007)
DHHS	Bioterrorism: CDC Intramural Research	2006	Decrease annual costs for personnel and materials development with development and continuous improvement of budget and performance integration information system tools (OMB 2007)
DHHS	NIOSH	2004	Percentage of grant award or funding decisions made available to applicants within 9 months of application receipt or deadline date while maintaining credible and efficient two-level peer-review system (OMB 2007)
DHHS	NIOSH	Not used currently	Determine future human capital resources needed to support programmatic strategic goals, focusing on workforce development or training and succession planning (Sinclair 2007)

DHHS	NIOSH	2007	Percentage of grant award or funding decisions made available to applicants within 9 months of application receipt or deadline date while maintaining credible and efficient two-level peer-review system (Sinclair 2007)
DHHS	Extramural Construction		By 2010, achieve average annual cost savings of managing construction grants by expanding use of electronic project-management tools that enhance oversight and 20-year use monitoring
			(Each FY) Achieve average annual cost of managing construction grants (Duran 2007)
DHHS	HIV/AIDS Research		By 2010, use enhanced AIDS Research Information System (ARIS) database to more efficiently conduct portfolio analysis to invest in priority AIDS research
			(FY 2005) Improve existing ARIS by converting its mainframe system into Web-based system designed by OAR and IC representatives in consultation with a contractor
			(FY 2006, FY 2007, FY 2008) Track, monitor, and budget for trans-NIH AIDS research, using enhanced ARIS database, to more efficiently conduct portfolio analysis of 100% of expiring grants to determine reallocation of resources for priority research (Duran 2007)
DHHS	Research Training Program	2006	By 2012, ensure that 100% of trainee appointment forms are processed electronically, to enhance program management (OMB 2007)
NASA	Human Systems Research and Technology	2005	Time between solicitation and selection in NASA Research Announcements (OMB 2007)
NASA	Solar System Exploration	2006	Percentage of budget for research projects allocated through open peer-reviewed competition (OMB 2007)
NASA	Solar System Exploration	2006	Number of days within which NASA Research Announcement research grants for program are awarded, from proposal due date to selection, with goal of 130 days (OMB 2007)

(Continued)

TABLE E-1 Continued

Agency or Organization	Program	Year	Efficiency Measure
NASA	Original Uniform Measures		Complete all development projects within 110% of cost and schedule baseline
			Peer-review and competitively award at least 80%, by budget, of research projects
			Reduce time within which 80% of NRA research grants are awarded, from proposal due date to selection, by 5% per year, with goal of 130 days
			Deliver at least 90% of scheduled operating hours for all operations and research facilities (Pollitt 2007)
NASA		2007	Year-to-year reduction in Space Shuttle sustaining engineering workforce for flight hardware and software while maintaining safe flight
			Reduction in ground operations cost (through 2012) of Constellation Systems based on comparison with Space Shuttle Program
			Number of financial processing steps and time to perform year-end closing
			Number of hours required for NASA personnel to collect, combine, and reconcile data of contract-management type for external agency reporting purposes (Pollitt 2007)
NASA		2007	On-time availability and operation of Aeronautics Test Program ground test facilities in support of research, development, test, and engineering milestones of NASA and DOD programs from both schedule and cost perspectives
			Operational cost per minute of Space Network support of missions
			Ratio of Launch Services Program cost per mission to total spacecraft cost
			Number of people reached via e-education technologies per dollar invested (Pollitt 2007)

Agency	Program	Year	Measure
NOAA	Climate Program	2004	Volume of data taken in annually and placed into archive (terabytes) (EPA, unpublished material, 2006)
NOAA	Ecosystem Research	2005	Cost per site characterization (OMB 2007)
NOAA	Ecosystem Research	2005	Percentage of grants awarded on time (OMB 2007)
NSF	Fundamental Science and Engineering Research	2005	Percentage of award decisions made available to applicants within 6 months of proposal receipt or deadline date while maintaining credible and efficient competitive merit-review system as evaluated by external experts (OMB 2007)
NSF	Research on Biocomplexity in the Environment	2004	Percentage of award decisions made available to applicants within 6 months of proposal receipt or deadline date while maintaining credible and efficient competitive merit-review system as evaluated by external experts (OMB 2007)
NSF	Construction and Operations of Research Facilities	2003	Percentage of construction acquisition and upgrade projects with negative cost and schedule variances of less than 10% of approved project plan (EPA, unpublished material, 2006)
NSF	Polar Research Tools, Facilities and Logistics	2004	Percentage of construction cost and schedule variances of major projects as monitored by earned-value management (OMB 2007)
NSF	Support for Research Institutions	2004	Percentage of award decisions made available to applicants within 6 months of proposal receipt or deadline date while maintaining credible and efficient competitive merit-review system as evaluated by external experts (OMB 2007)
NSF	Support for Small Research Collaborations	2004	Percentage of award decisions made available to applicants within 6 months of proposal receipt or deadline date while maintaining credible and efficient competitive merit-review system as evaluated by external experts (OMB 2007)
NSF	Construction and Operations of Research Facilities	2003	Percentage of operational facilities that keep scheduled operating time lost to less than 10% (OMB 2007)
NSF	Federally Funded Research and Development Centers	2005	Percentage of operational facilities that keep scheduled operating time lost to less than 10% (OMB 2007)

(Continued)

TABLE E-1 Continued

Agency or Organization	Program	Year	Efficiency Measure
NSF	Information Technology Research		Qualitative assessment by external experts that there have been significant research contributions to software design and quality, scalable information infrastructure, high-end computing, workforce, and socioeconomic impacts of IT (EPA, unpublished material, 2006)
NSF	Polar Research Tools, Facilities and Logistics		Percentage of person-days planned for Antarctic research for which program is able to provide necessary research support (EPA, unpublished material, 2006)
NSF	Polar Research Facilities and Support		Research facilities: keep construction cost and schedule variances of major polar facilities projects as monitored by earned-value management at 8% or less Research support: provide necessary research support for Antarctic researchers at least 90% of time (OMB 2007)
NSF	Support for Individual Researchers		External validation of "significant achievement" in attracting and preparing U.S. students to be highly qualified members of global S&E workforce (EPA, unpublished material, 2006)
NSF	Science and Engineering Centers Program	2006	Percentage of decisions on preproposals that are merit-reviewed and available to Centers Program applicants within 5 months of preproposal receipt or deadline date (OMB 2007)
NSF			Time to decision for proposals: for 70% of proposals submitted to National Science Foundation, inform applicants about funding decisions within 6 months of proposal receipt or deadline date or target date, whichever is later (Tsuchitani 2007)
NSF			Facilities cost, schedule, and operations: keep negative cost and schedule variances at less than 10% of approved project plan for 90% of facilities; keep loss of operating time due to unscheduled downtime to less than 10% of total scheduled operating time for 90% of operational facilities (Tsuchitani 2007)

USDA	USDA Research: Economic Opportunities for Producers		Percentage of construction acquisition and upgrade projects with negative cost variance of less than 10% of approved project plan (EPA, unpublished material, 2006)
USDA	Economic Opportunities for Producers	2004	Cumulative dollars saved for grant review (OMB 2007)
USDA	Economic Opportunities for Producers	2004	Proposal review time in days (OMB 2007)
USDA	Research on Protection and Safety of Agricultural Food Supply	2005	Additional research funds leveraged from external sources (OMB 2007)
USDA	Economic Research Service	2005	Index of ERS product releases per staff year (OMB 2007)
USDA	Grants for Economic Opportunities and Quality of Life for Rural America	2006	Cumulative dollars saved for grant review: dollars saved reflect average salary saved by calculating number of calendar days saved annually between receipt of proposal and date funding awarded for competitively reviewed proposals, then multiplied by average daily salary for CSREES employees (OMB 2007)
USDA	In-House Research for Natural Resource Base and Environment	2006	Relative increase in peer-reviewed publications (OMB 2007)
USDA	In-House Research for Nutrition and Health	2006	Relative increase in peer-reviewed publications (OMB 2007)

(Continued)

TABLE E-1 Continued

Agency or Organization	Program	Year	Efficiency Measure
Alcoa			Return-on-investment calculation: (FY 2005) Improve existing ARIS by converting its mainframe system into Web-based system designed by OAR and IC representatives in consultation with contractor
			Variable cost improvement
			Margin impact from organic growth
			Capital avoidance
			Cost avoidance
			Annual impact of these four metrics over 5-year period becomes numerator; denominator is total R&D budget
			Metric is used most often to evaluate overall value of R&D program and current budget focus (Atkins 2007)
Alcoa			Time (Atkins 2007)
Alcoa			Cost (Atkins 2007)
Alcoa			Customer demand (Atkins 2007)
Alcoa			Risk (Atkins 2007)
Alcoa			Impact on business (Atkins 2007)
Alcoa			Impact on customers (Atkins 2007)
Alcoa			Location (Atkins 2007)
Alcoa			Intellectual property (Atkins 2007)
Alcoa			Aggregate R&D expenditures by laboratory group or by identifiable programs and publish value capture or "success rate" for each on annual basis (Atkins 2007)
Alcoa			ROI on R&D spending; success rate of launched products (Atkins 2007)

Dow Chemical	Publications; participation and leadership in scientific community (collaborative research efforts; trade associations; ILSI-HESI; external workshops; adjunct faculty positions, journal or book editors, professional societies) (Bus 2007)
IBM	ROI on Summer Internship Program and Graduate Fellowship Program: what percentage return as regular IBM research employees?
IBM	"Bureaucracy Busters" Initiative to reduce bureaucracy in laboratory support, information-technology support, HR processes, and business processes (Kenney 2007)
IBM	Tracking of patent-evaluation process (Kenney 2007)
IBM	Customer-satisfaction surveys for support functions to evaluate effect of service reductions (Kenney 2007)
IBM	Measurement of response time and turnaround for external contracts (Kenney 2007)
IBM	Measurement of span of responsibility for secretarial support (Kenney 2007)
Procter & Gamble	Time saved in product development (Daston 2007)
Procter & Gamble	Increased confidence about safety (Daston 2007)
Procter & Gamble	External relations benefits (although not quantifiable) (Daston 2007)

REFERENCES

Atkins, P. 2007. Alcoa Research and Development. Presentation at the Workshop on Evaluating the Efficiency of Research and Development Programs at the Environmental Protection Agency, April 24, 2007, Washington, DC.

Beschen, D. 2007. Federal Research Agencies' Overview. Presentation at the Workshop on Evaluating the Efficiency of Research and Development Programs at the Environmental Protection Agency, April 24, 2007, Washington, DC.

Bus, J.S. 2007. Research Efficiency: Industry Perspective. Presentation at the Workshop on Evaluating the Efficiency of Research and Development Programs at the Environmental Protection Agency, April 24, 2007, Washington, DC.

Daston, G. 2007. Evaluating Efficiency of P&G Central Product Safety Research. Presentation at the Workshop on Evaluating the Efficiency of Research and Development Programs at the Environmental Protection Agency, April 24, 2007, Washington, DC.

Duran, D. 2007. Measuring Efficiency in Science. Presentation at the Workshop on Evaluating the Efficiency of Research and Development Programs at the Environmental Protection Agency, April 24, 2007, Washington, DC.

Kenney, J. 2007. Evaluating R&D Efficiency IBM's Perspective-Within and Without IBM. Presentation at the Workshop on Evaluating the Efficiency of Research and Development Programs at the Environmental Protection Agency, April 24, 2007, Washington, DC.

OMB (Office of Management and Budget). 2007. ExpectMore.gov. Office of Management and Budget [online]. Available: http://www.whitehouse.gov/omb/expect more/ [accessed April 5, 2007].

Pollitt, J.A. 2007. Evaluating the Efficiency of NASA's R&D Programs. Presentation at the Workshop on Evaluating the Efficiency of Research and Development Programs at the Environmental Protection Agency, April 24, 2007, Washington, DC.

Sinclair, R. 2007. PART Efficiency Measures at NIOSH. Presentation at the Workshop on Evaluating the Efficiency of Research and Development Programs at the Environmental Protection Agency, April 24, 2007, Washington, DC.

Tsuchitani, P. 2007. PART Efficiency Measures at NIOSH. Presentation at the Workshop on Evaluating the Efficiency of Research and Development Programs at the Environmental Protection Agency, April 24, 2007, Washington, DC.

Appendix F

Draft Board of Scientific Counselors Handbook for Subcommittee Chairs: Draft Proposed Charge Questions for BOSC Reviews[1]

PROGRAM ASSESSMENT
(EVALUATE ENTIRE RESEARCH PROGRAM)

The responses to the program assessment charge questions below will be in a narrative format, and will capture the performance for the *entire* research program and all the activities in support of the program's Long Term Goals (LTGs). The Long term Goals should be consistent with EPA's Strategic Plan and mutually agreed upon by ORD and OMB.

Program Relevance

1. How consistent are the Long Term Goals (LTGs) of the program with achieving the Agency's strategic plan and ORD's Multi-Year Plan?

2. How responsive is the program focus to program office and regional research needs?

3. How responsive is the program to recommendations from outside advisory boards and stakeholders?

4. How clearly evident are the public benefits of the program?

Program Structure

1. How clear a logical framework do the LTGs provide for organizing and planning the research and demonstrating outcomes of the program?

---

[1]EPA 2007.

2. How appropriate is the science used to achieve each LTG, i.e., is the program still asking the right questions, or has it been eclipsed by advancements in the field?

3. Does the MYP describe an appropriate flow of work (i.e., the sequencing of related activities) that reasonably reflects the anticipated pace of scientific progress and timing of client needs?

4. Does the program use the MYP to help guide and manage its research?

5. How logical is the program design, with clearly identified priorities?

Program Performance

1. How much progress is the program making on each LTG based on clearly stated and appropriate milestones?

Program Quality

1. How good is the scientific quality of the program's research products?

2. What means does the program employ to ensure quality research (including peer review, competitive funding, etc.?

3. How effective are these processes?

Scientific Leadership

1. Please comment on the leadership role the research program and its staff have in contributing to advancing the current state of the science and solving important research problems.

Coordination and Communication

1. How effectively does the program engage scientists and managers from ORD and relevant program offices in its planning?

2. How effectively does the program engage outside organizations, both within and outside government, to promote collaboration, obtain input on program goals and research, and avoid duplication of effort?

3. How effective are the mechanisms that the program uses for communicating research results both internally and externally?

Outcomes

1. How well-defined are the program's measures of outcomes?

2. How much are the program results being used by environmental decision makers to inform decisions and achieve results?

SUMMARY ASSESSMENT
(RATE PROGRAM PERFORMANCE BY LTG)

The responses to the three summary assessment charge questions below will rate the performance for each LTG. For each LTG, a qualitative score will be assigned that reflects the quality and significance of the research as well as the extent to which the program is meeting or making measurable progress toward the goal—relative to the information and evidence provided to the BOSC. The scores will be given in the form of adjectives that are clearly defined and which are intended to promote consistency among reviews. The adjectives will be used as part of a narrative summary of the review of each LTG so that the context of the rating and the rationale for selecting a particular rating will be transparent. The rating may reflect considerations beyond the summary assessment questions, and will be explained in the narrative. The adjectives to describe progress are:

- *Exceptional*: indicates that the program is meeting all and exceeding some of its goals, both in the quality of the science being produced and the speed at which research result tools and methods are being produced. An exceptional rating also indicates that the program is addressing the right questions to achieve its goals. The review should be specific as to which aspects of the program's performance have been exceptional.
- *Exceeds Expectations*: indicates that the program is meeting all of its goals. It addresses the appropriate scientific questions to meet its goals and the science is competent or better. It exceeds expectations for either the high quality of the science *or* for the speed at which work products are being produced and milestones met.
- *Meets Expectations*: indicates that the program is meeting most of its goals. Satisfactory programs live up to expectations in terms of addressing the appropriate scientific questions to meet its goals, and that work products are being produced and milestones are being reached in a timely manner. The quality of the science being done is competent or better.
- *Not Satisfactory*: indicates that the program is failing to meet a substantial fraction of its goals, or if meeting them, that the achievement of milestones is significantly delayed, or that the questions being addressed are inappropriate or insufficient to meet the intended purpose. Questionable science is also a reason for rating a program as unsatisfactory for a particular long term goal. The review should be specific as to which aspects of a program's performance have been inadequate.

For each program review, the summary assessment charge questions below will be tailored to the specific review and LTG:

1. How appropriate is the science used to achieve each LTG, i.e., is the program still asking the right questions, or has it been eclipsed by advancements in the field?

2. How good is the scientific quality of the program's research products?

3. How much are the program results being used by environmental decision makers to inform decisions and achieve results?

REFERENCES

EPA (U.S. Environmental Protection Agency). 2007. Draft Board of Scientific Counselors Handbook for Subcommittee Chairs. Board of Scientific Counselors, U.S. Environmental Protection Agency, Washington, DC.

Appendix G

OMB's Research and Development Program Investment Criteria[1,2]

As another initiative of the President's Management Agenda, the development of explicit R&D investment criteria builds on the best of the planning and assessment practices that R&D program managers use to plan and assess their programs. The Administration has worked with experts and stakeholders to build upon lessons learned from previous approaches.

Agencies should use the criteria as broad guidelines that apply at all levels of Federally funded R&D efforts, and they should use the PART as the instrument to periodically evaluate compliance with the criteria at the program level. To make this possible, the R&D PART aligns with the R&D criteria. The R&D criteria are reprinted here as a guiding framework for addressing the R&D PART.

The R&D criteria address not only planning, management, and prospective assessment but also retrospective assessment. Retrospective review of whether investments were well-directed, efficient, and productive is essential for validating program design and instilling confidence that future investments will be wisely invested. Retrospective reviews should address continuing program relevance, quality, and successful performance to date.

While the criteria are intended to apply to all types of R&D, the Administration is aware that predicting and assessing the outcomes of *basic* research in particular is never easy. Serendipitous results are often the most interesting and

[1](OMB 2007).

[2]To assist agencies with significant research programs, additional instructions were added to the PART Guidance and titled the "Research and Development Program Investment Criteria." The R&D Investment Criteria are found in Appendix C of the PART instructions. Unlike the main body of the PART instructions, which apply to all federal agencies and programs, the R&D Investment Criteria attempt to clarify OMB's expectations specifically for R&D programs.

ultimately may have the most value. Taking risks and working toward difficult-to-attain goals are important aspects of good research management, and innovation and breakthroughs are among the results. However, there is no inherent conflict between these facts and a call for clearer information about program goals and performance toward achieving those goals. The Administration expects agencies to focus on improving the management of their research programs and adopting effective practices, and not on predicting the unpredictable.

The R&D investment criteria have several potential benefits:

- Use of the criteria allows policy makers to make decisions about programs based on information beyond anecdotes, prior-year funding levels, and lobbying of special interests.
- A dedicated effort to improve the process for budgeting, selecting, and managing R&D programs is helping to increase the return on taxpayer investment and the productivity of the Federal R&D portfolio.
- The R&D investment criteria will help communicate the Administration's expectations for proper program management.
- The criteria and subsequent implementation guidance will also set standards for information to be provided in program plans and budget justifications.
- The processes and collected information promoted under the criteria will improve public understanding of the possible benefits and effectiveness of the Federal investment in R&D.

DETAILS ON THE CRITERIA

The Relevance, Quality, and Performance criteria apply to all R&D programs. Industry- or market-relevant applied R&D must meet additional criteria. Together, these criteria can be used to assess the need, relevance, appropriateness, quality, and performance of Federal R&D programs.

Relevance

R&D investments must have clear plans, must be relevant to national priorities, agency missions, relevant fields, and "customer" needs, and must justify their claim on taxpayer resources. Programs that directly support Presidential priorities may receive special consideration with adequate documentation of their relevance. Review committees should assess program objectives and goals on their relevance to national needs, "customer" needs, agency missions, and the field(s) of study the program strives to address. For example, the Joint DOE/NSF Nuclear Sciences Advisory Committee's Long Range Plan and the Astronomy Decadal Surveys are the products of good planning processes because they articulate goals and priorities for research opportunities within and across their respective fields.

OMB will work with some programs to identify quantitative metrics to estimate and compare potential benefits across programs with similar goals. Such comparisons may be within an agency or among agencies.

Programs Must Have Complete Plans, With Clear Goals and Priorities

Programs must provide complete plans, which include explicit statements of:

- specific issues motivating the program;
- broad goals and more specific tasks meant to address the issues;
- priorities among goals and activities within the program;
- human and capital resources anticipated; and
- intended program outcomes, against which success may later be assessed.

Programs Must Articulate the Potential Public Benefits of the Program

Programs must identify potential benefits, including added benefits beyond those of any similar efforts that have been or are being funded by the government or others. R&D benefits may include technologies and methods that could provide new options in the future, if the landscape of today's needs and capabilities changes dramatically. Some programs and sub-program units may be required to quantitatively estimate expected benefits, which would include metrics to permit meaningful comparisons among programs that promise similar benefits. While all programs should try to articulate potential benefits, OMB and OSTP recognize the difficulty in predicting the outcomes of basic research. Consequently, agencies may be allowed to relax this as a requirement for basic research programs.

Programs Must Document Their Relevance to Specific Presidential Priorities to Receive Special Consideration

Many areas of research warrant some level of Federal funding. Nonetheless, the President has identified a few specific areas of research that are particularly important. To the extent a proposed project can document how it directly addresses one of these areas, it may be given preferential treatment.

Program Relevance to the Needs of the Nation, of Fields of Science and Technology [S&T], and of Program "Customers" Must Be Assessed Through Prospective External Review

Programs must be assessed on their relevance to agency missions, fields of science or technology, or other "customer" needs. A customer may be another program at the same or another agency, an interagency initiative or partnership, or a firm or other organization from another sector or country. As appropriate, programs must define a plan for regular reviews by primary customers of the program's relevance to their needs. These programs must provide a plan for addressing the conclusions of external reviews.

Program Relevance to the Needs of the Nation, of Fields of S&T, and of Program "Customers" Must Be Assessed Periodically Through Retrospective External Review

Programs must periodically assess the need for the program and its relevance to customers against the original justifications. Programs must provide a plan for addressing the conclusions of external reviews.

Quality

Programs should maximize the quality of the R&D they fund through the use of a clearly stated, defensible method for awarding a significant majority of their funding. A customary method for promoting R&D quality is the use of a competitive, merit-based process. NSF's process for the peer-reviewed, competitive award of its R&D grants is a good example. Justifications for processes other than competitive merit review may include "outside-the-box" thinking, a need for timeliness (e.g., R&D grants for rapid response studies of *Pfisteria*), unique skills or facilities, or a proven record of outstanding performance (e.g., performance-based renewals).

Programs must assess and report on the quality of current and past R&D. For example, NSF's use of Committees of Visitors, which review NSF directorates, is an example of a good quality-assessment tool. OMB and OSTP encourage agencies to provide the means by which their programs may be benchmarked internationally or across agencies, which provides one indicator of program quality.

Programs Allocating Funds Through Means Other Than a Competitive, Merit-based Process Must Justify Funding Methods and Document How Quality is Maintained

Programs must clearly describe how much of the requested funding will be broadly competitive based on merit, providing compelling justifications for R&D funding allocated through other means. (See OMB Circular A-11 for definitions of competitive merit review and other means of allocating Federal research funding.) All program funds allocated through means other than unlimited competition must document the processes they will use to distribute funds to each type of R&D performer (e.g., Federal laboratories, Federally-funded R&D

centers, universities, etc.). Programs are encouraged to use external assessment of the methods they use to allocate R&D and maintain program quality.

Program Quality Must Be Assessed Periodically Through Retrospective Expert Review

Programs must institute a plan for regular, external reviews of the quality of the program's research and research performers, including a plan to use the results from these reviews to guide future program decisions. Rolling reviews performed every 3-5 years by advisory committees can satisfy this requirement. Benchmarking of scientific leadership and other factors provides an effective means of assessing program quality relative to other programs, other agencies, and other countries.

Performance

R&D programs should maintain a set of high priority, multi-year R&D objectives with annual performance outputs and milestones that show how one or more outcomes will be reached. Metrics should be defined not only to encourage individual program performance but also to promote, as appropriate, broader goals, such as innovation, cooperation, education, and dissemination of knowledge, applications, or tools.

OMB encourages agencies to make the processes they use to satisfy the Government Performance and Results Act (GRPA) consistent with the goals and metrics they use to satisfy these R&D criteria. Satisfying the R&D performance criteria for a given program should serve to set and evaluate R&D performance goals for the purposes of GPRA. OMB expects goals and performance measures that satisfy the R&D criteria to be reflected in agency performance plans.

Programs must demonstrate an ability to manage in a manner that produces identifiable results. At the same time, taking risks and working toward difficult-to-attain goals are important aspects of good research management, especially for basic research. The intent of the investment criteria is not to drive basic research programs to pursue less risky research that has a greater chance of success. Instead, the Administration will focus on improving the management of basic research programs.

OMB will work with some programs to identify quantitative metrics to compare performance across programs with similar goals. Such comparisons may be within an agency or among agencies.

Construction projects and facility operations will require additional performance metrics. Cost and schedule earned-value metrics for the construction of R&D facilities must be tracked and reported. Within DOE, the Office of Science's formalized independent reviews of technical cost, scope, and schedule baselines and project management of construction projects ("Lehman Reviews") are widely recognized as an effective practice for discovering and correcting problems involved with complex, one-of-a-kind construction projects.

REFERENCES

OMB (Office of Management and Budget). 2007. Research and development program investment criteria. Pp. 72-77 in Guide to the Program Assessment Rating Tool (PART). Program Assessment Rating Tool Guidance No. 2007-02. Office of Management and Budget, Washington, DC. January 29, 2007 [online]. Available: http://stinet.dtic.mil/cgi-bin/GetTRDoc?AD=ADA471562&Location=U2&doc=GetTRDoc.pdf [accessed Nov. 14, 2007].

Appendix H

Charge to the BOSC Subcommittee on Safe Pesticides/Safe Products Research[1]

OBJECTIVE

The BOSC Safe Pesticides/Safe Products (SP2) Subcommittee will conduct a retrospective and prospective review of ORD's SP2 Research Program, and evaluate the program's relevance, quality, performance, and scientific leadership. The BOSC's evaluation and recommendations will provide guidance to the Office of Research and Development to help:

- plan, implement, and strengthen the program;
- compare the program with programs designed to achieve similar outcomes in other parts of EPA and in other federal agencies;
- make research investment decisions over the next five years;
- prepare EPA's performance and accountability reports to Congress under the Government Performance and Results Act; and
- respond to assessments of federal research programs such as those conducted by the

Office of Management and Budget (OMB highlights the value of recommendations from independent expert panels in guidance to federal agencies).

BACKGROUND INFORMATION

Independent expert review is used extensively in industry, federal agencies, Congressional committees, and academia. The National Academy of Science has recommended this approach for evaluating federal research programs.

---

[1](EPA 2007).

Because of the nature of research, it is not possible to measure the creation of new knowledge as it develops–or the pace at which research progresses or scientific breakthroughs occur. Demonstrating research contributions to outcomes is very challenging when federal agencies conduct research to support regulatory decisions, and then rely on third parties–such as state environmental agencies–to enforce the regulations and demonstrate environmental improvements. Typically, many years may be required for practical research applications to be developed and decades may be required for some research outcomes to be achieved in a measurable way.

Most of ORD's environmental research programs investigate complex environmental problems and processes—combining use-inspired basic research with applied research, and integrating several scientific disciplines across a conceptual framework that links research to environmental decisions or environmental outcomes. In multidisciplinary research programs such as these, progress toward outcomes can not be measured by outputs created in a single year. Rather, research progress occurs over several years, as research teams explore hypotheses with individual studies, interpret research findings, and then develop hypotheses for future studies.

In designing and managing its research programs, ORD emphasizes the importance of identifying priority research questions or topics to guide its research. Similarly, ORD recommends that its programs develop a small number of performance goals that serve as indicators of progress to answer the priority questions and to accomplish outcomes. Short-term outcomes are accomplished when research is applied by specific clients, e.g., to strengthen environmental decisions. These decisions and resulting actions (e.g., the reduction of contaminant emissions or restoration of ecosystems) ultimately contribute to improved environmental quality and health.

In a comprehensive evaluation of science and research at EPA, the National Research Council recommended that the Agency substantially increase its efforts to both explain the significance of its research products and to assist clients inside and outside the Agency in applying them. In response to this recommendation, ORD has engaged science advisors from client organizations to serve as members of its research program teams. These teams help identify research contributions with significant decision making value and help plan for their transfer and application.

For ORD's environmental research programs, periodic retrospective analysis at intervals of four or five years is needed to characterize research progress, to assess how clients are applying research to strengthen environmental decisions, and to evaluate client feedback about the research. Conducting program evaluations at this interval enables assessment of: research progress, the scientific quality and decision-making value of the research, and whether research progress has resulted in short-term outcomes for specific clients.

A description of the OSTP/OMB *Research and Development Investment Criteria* is included in Appendix I.

BACKGROUND FOR ORD'S SP2 RESEARCH PROGRAM
AND DRAFT CHARGE QUESTIONS BACKGROUND

The purpose of the SP2 Research Program is to provide EPA's Office of Prevention, Pesticides, and Toxic Substances (OPPTS) with the scientific information it needs to reduce or prevent unreasonable risks to humans, wildlife, and non-target plants from exposures to pesticides, toxic chemicals, and products of biotechnology. The SP2 Research Program specifically addresses OPPTS' high priority research needs that are not addressed by any of ORD's other research programs. The research program is focused on three Long Term Goals:

Long Term Goal 1: OPPTS and/or other organizations use the results of ORD's research on methods, models, and data as the scientific foundation for: A) prioritization of testing requirements, B) enhanced interpretation of data to improve human health and ecological risk assessments, and C) decisionmaking regarding specific individual or classes of pesticides and toxic substances that are of high priority. *The ultimate outcomes are the development of improved methods, models, and data for OPPTS' use in requiring testing, evaluating data, completing risk assessments, and determining risk management approaches. More specifically the outcomes are the development by ORD and implementation by OPPTS of more efficient and effective testing paradigms that will be better informed by predictive tools (chemical identification, improved targeting, less cost, less time, and fewer animals); improved methods by which data from the more efficient and effective testing paradigms can be integrated into risk assessments; and that OPPTS uses the result of ORD's multidisciplinary research approaches, that it specifically requests, for near term decisionmaking on high priority individual or classes of pesticides and toxic substances.*

Long Term Goal 2: OPPTS and/or other organizations use the results of ORD's research as the scientific foundation for probabilistic risk assessments to protect natural populations of birds, fish, other wildlife, and non-target plants. *Results of this research will help the Agency meet the long term goal of developing scientifically valid approaches to extrapolate across species, biological endpoints and exposure scenarios of concern, and to assess spatially explicit, population-level risks to wildlife populations and non-target plants and plant communities from pesticides, toxic chemicals and multiple stressors, while advancing the development of probabilistic risk assessment.*

Long Term Goal 3: OPPTS and/or other organizations use the results of ORD's biotechnology research as the scientific foundation for decisionmaking related to products of biotechnology. *OPPTS will use the results from this research program to update its requirements of registrants of products of biotechnology and to help evaluate data submitted for its review.*

The scope of the SP2 research program has been developed in partnership with OPPTS. ORD keeps abreast of complementary research ongoing in other federal agencies and scientific organizations. However, no other programs have similar goals, in terms of scope and mission, as the SP2 research program that provides OPPTS with the tools it needs to carry out its regulatory mandates. EPA's SP2 research is multi-disciplinary, including: 1) research across all aspects of the risk assessment/risk management paradigm, i.e., in effects, exposure, risk assessment, and risk management; and 2) as related to humans, wildlife, and plants. Comparison of potential benefits is conducted from a scientific perspective through coordinating and collaborating with other research programs, participating at national and international scientific for a, and keeping abreast of state of the science. EPA's SP2 program includes many areas that are of unique importance in helping OPPTS meet its legislative mandates, such as requiring industry to submit data on pesticides, toxic substances, and products of biotechnology. The SP2 program also includes other research areas that serve to improve the basic scientific understanding regarding these agents that OPPTS and other parts of the Agency need to evaluate data submissions, conduct risk assessments, and make informed management decisions. Furthermore, ORD's intramural program is complemented by an extramural program implemented through the Science to Achieve Results (STAR) program.

The research directions to address the key areas of scientific uncertainty are captured in the current version of the SP2 Multi-Year Plan (MYP). The MYP includes research activities implemented and planned for the period 2007 through 2015. The research described in the MYP assumes annual intramural and extramural resources of approximately 126 FTEs and $24.8 million, including payroll, travel and operating expenses.

DRAFT CHARGE

Program Assessment (Evaluate Entire Research Program)

The responses to the program assessment charge questions below should be in a narrative format, and should capture the performance for the entire research program and all the activities in support of the program's Long Term Goals (LTGs).

Program Relevance

1. How consistent are the Long Term Goals (LTGs) of the program with achieving the Agency's strategic plan and ORD's Multi-Year Plan?
2. How responsive is the program focus to program office and regional research needs?

3. How responsive is the program to recommendations from outside advisory boards and stakeholders?

4. How clearly evident are the public benefits of the program?

Factors to consider: the degree to which the research is driven by EPA priorities; the degree to which this research program has had (or is likely to have) an impact on Agency decisionmaking; and the extent to which research program scientists participate on or contribute to Agency workgroups and transfer research to program and regional customers.

Program Structure

1. How clear a logical framework do the LTGs provide for organizing and planning the research and demonstrating outcomes of the program?

2. How appropriate is the science used to achieve each LTG, i.e., is the program asking the right questions, or has it been eclipsed by advancements in the field?

3. Does the MYP describe an appropriate flow of work (i.e., the sequencing of related activities) that reasonably reflects the anticipated pace of scientific progress and timing of client needs?

4. Does the program use the MYP to help guide and manage its research?

5. How logical is the program design, with clearly identified priorities?

Factors to consider: the appropriateness of the key science questions; the appropriateness of the Long Term Goals in providing a logical framework for organizing the SP2 program to best meet the Agency's needs; the degree of clarity to the path of annual research products aimed at accomplishing each of the LTGs; the scientific soundness of the approaches used; the appropriateness of the research products identified in the MYP as the means to meet the highest priority research for each LTG; and the adequacy/sufficiency/necessity of the sets of APMs under the APGs to accomplish the intended goals.

Program Performance

1. How much progress is the program making on each LTG based on clearly stated and appropriate milestones?

Factors to consider: the scientific soundness of the approaches used; the degree to which scientific understanding of the problem has been advanced; the degree to which scientific uncertainty has been reduced; the impact and use of research results by EPA program and regional offices and by other organizations; and the extent of the bibliography of peer reviewed publications.

Program Quality

1. How good is the scientific quality of the program's research products?
2. What means does the program employ to ensure quality research (including peer review, competitive funding, etc.)?
3. How effective are these processes?

Factors to consider: the impact and use of research results by EPA program and regional offices and other organizations; the degree to which peer reviewed publications from this program are cited in other peer reviewed publications, the immediacy with which they are cited, and their impact factor; the processes used to peer review intramural research designs and products (e.g., division-level or product-level reviews by independent panels); and the processes used in the competitive extramural grants program.

Scientific Leadership

1. Please comment on the leadership role the research program and its staff have in contributing to advancing the current state of the science and solving important research problems.

Factors to consider: the degree to which this program is identified as a leader in the field; the degree to which peer reviewed publications from this program are cited in other peer reviewed publications, the immediacy with which they are cited, and their impact factor; the degree to which SP2 scientists serve/are asked to serve on national/international workgroups, officers in professional societies, publication boards; the degree to which SP2 scientists lead national/international collaborative efforts, organize national/international conferences/symposia, and are awarded for their contributions/leadership; and benchmarking of scientific leadership relative to other programs, agencies, and countries.

Coordination and Communication

1. How effectively does the program engage scientists and managers from ORD and relevant program offices in its planning?
2. How effectively does the program engage outside organizations, both within and outside government, to promote collaboration, obtain input on program goals and research, and avoid duplication of effort?
3. How effective are the mechanisms that the program uses for communicating research results both internally and externally?

Factors to consider: the extent to which program/regional office scientists/managers are involved in planning the research; research activities of other federal agencies, industry, academic institutions, other countries; the degree of collaboration and coordination with other research organizations; and the means that are used to communicate results to OPPTS and to the external scientific community (e.g., through peer reviewed publications, scientific meetings, seminars).

Outcomes

1. How well-defined are the program's measures of outcomes?
2. How much are the program results being used by environmental decision makers to inform decisions and achieve results?

Factors to consider: the extent to which the MYP identifies the past or anticipated impact of the research activities; and the extent to which the research has contributed/or is anticipated to contribute to Agency and other decision-making.

Summary Assessment (Rate Program Performance By LTG)

A summary assessment and narrative should be provided for each LTG. The assessment should be based on 3 of the questions included above, which are:

1. How appropriate is the science used to achieve each LTG, i.e., is the program asking the right questions, or has it been eclipsed by advancements in the field?
2. How good is the scientific quality of the program's research products?
3. How much are the program results being used by environmental decision makers to inform decisions and achieve results?

Elements to Include for Long-Term Goal 1

The appropriateness, quality, and use of ORD science by OPPTS and other organizations to inform decisions and achieve results with respect to 1) prioritization testing requirements, 2) enhancing the interpretation of data to improve human health and ecological risk assessments, and 3) making decisions regarding specific individual or classes of high priority pesticides and toxic substances. The extent to which ORD is asking the right questions, conducting the right science, and providing products that are responsive to OPPTS's and other organizations' needs.

Elements to Include for Long-Term Goal 2

The appropriateness, quality, and use of ORD science by OPPTS and other organizations to inform decisions and achieve results with respect to probabilistic risk assessments to protect natural populations of birds, fish, other wildlife, and non-target plants. The extent to which ORD is asking the right questions, conducting the right science, and providing products that are responsive to OPPTS' and other organizations' needs

Elements to Include for Long-Term Goal 3

The appropriateness, quality, and use of ORD science by OPPTS and other organizations to inform decisions and achieve results with respect to products of biotechnology. The extent to which ORD is asking the right questions, conducting the right science, and providing products that are responsive to OPPTS' and other organizations' needs.

For each LTG, the BOSC SP2 Subcommittee will assign a qualitative score that reflects the quality and significance of the research as well as the extent to which the program is meeting or making measurable progress toward the goal—relative to the evidence provided to the BOSC. The scores should be in the form of the following adjectives that are defined below and intended to promote consistency among BOSC program reviews. The adjectives should be used as part of a narrative summary of the review, so that the context of the rating and the rationale for selecting a particular rating will be transparent. The rating may reflect considerations beyond the summary assessment questions, and will be explained in the narrative. The adjectives to describe progress are:

- Exceptional: indicates that the program is meeting all and exceeding some of its goals, both in the quality of the science being produced and the speed at which research result tools and methods are being produced. An exceptional rating also indicates that the program is addressing the right questions to achieve its goals. The review should be specific as to which aspects of the program's performance have been exceptional.

- Exceeds Expectations: indicates that the program is meeting all of its goals. It addresses the appropriate scientific questions to meet its goals and the science is competent or better. It exceeds expectations for either the high quality of the science or for the speed at which work products are being produced and milestones met.

- Meets Expectations: indicates that the program is meeting most of its goals. Programs meet expectations in terms of addressing the appropriate scientific questions to meet its goals, and that work products are being produced and milestones are being reached in a timely manner. The quality of the science being done is competent or better.

- Not Satisfactory: indicates that the program is failing to meet a substantial fraction of its goals, or if meeting them, that the achievement of milestones is significantly delayed, or that the questions being addressed are inappropriate or insufficient to meet the intended purpose. Questionable science is also a reason for rating a program as unsatisfactory for a particular long term goal. The review should be specific as to which aspects of a program's performance have been inadequate.

REFERENCES

EPA (U.S. Environmental Protection Agency). 2007. Review of the Office of Research and Development's Safe Pesticides/Safe Products (SP2) Research at the U.S. Environmental Protection Agency. Board of Scientific Counselors, U.S. Environmental Protection Agency, Washington, DC.

Appendix I

PART Guidance on Efficiency Measures[1]

DESCRIPTION OF EFFICIENCY MEASURES FOR PART

Efficiency Measures

While outcome measures provide valuable insight into program achievement, more of an outcome can be achieved with the same resources if an effective program increases its efficiency. The President's Management Agenda (PMA) Budget and Performance Integration (BPI) Initiative encourages agencies to develop efficiency measures. Sound efficiency measures capture skillfulness in executing programs, implementing activities, and achieving results, while avoiding wasted resources, effort, time, and/or money. Simply put, efficiency is the ratio of the outcome or output to the input of any program. Because they relate to costs, efficiency measures are likely to be annual measures.

- *Outcome efficiency measures:* The best efficiency measures capture improvements in program outcomes for a given level of resource use. Outcome efficiency measures are generally considered the best type of efficiency measure for assessing the program overall. For example, a program that has an outcome goal of "reduced energy consumption" may have an efficiency measure that shows the value of energy saved in relation to program costs.
- *Output efficiency measures:* It may be difficult to express efficiency measures in terms of outcomes. In such cases, acceptable efficiency measures could focus on how to produce a given output level with fewer resources. However, this approach should not shift incentives toward quick, low-quality methods that could hurt program effectiveness and desired outcomes.

---

[1]OMB 2006.

Meaningful efficiency measures consider the benefit to the customer and serve as indicators of how well the program performs. For example, reducing processing time means little if error rates increase. A balanced approach is required to enhance the performance of both variables in pursuit of excellence to customers. In these instances, one measure (e.g., increase in customer satisfaction) may be used in conjunction with another complementary measure (e.g., reduction in processing time).

In all cases, efficiency measures must be useful, relevant to program purpose, and help improve program performance. An efficiency measure for a Federal program tracks the ratio of total outputs or outcomes to total inputs (Federal plus non-Federal). Leveraging program resources can be a rational policy decision, as it leads to risk or cost sharing; however, it is not an acceptable efficiency measure, because the leveraging ratio of non-Federal to Federal dollars represents only inputs. Although increasing the amount leveraging in a program may stretch Federal program dollars, this does not measure improvements in the management of total program resources, systems, or outcomes.

3.4: Does the program have procedures (e.g., competitive sourcing/cost comparisons, IT improvements, appropriate incentives) to measure and achieve efficiencies and cost effectiveness in program execution?

Purpose: To determine whether the program has effective management procedures and measures in place to ensure the most efficient use of each dollar spent on program execution.

Elements of Yes: A *Yes* answer needs to clearly explain and provide evidence of each of the following [see Box I-1]:

- The program has regular procedures in place to achieve efficiencies and cost effectiveness.
- The program has at least one efficiency measure with baseline and targets.

BOX I-1 Measures and PARTWeb

To receive a *Yes* answer, the program must include at least one efficiency measure, baseline data/estimates, and targets in the Measures screen in PARTWeb.

Only measures that meet the standards for a *Yes* should be entered in PARTWeb.

Please ensure that the proper characterization of measures is selected in PARTWeb (that is "efficiency"). Make sure to indicate the term of the measure in PARTWeb too (that is, long-term, annual, or long-term/annual).

There are several ways to demonstrate that a program has established procedures for so improving efficiency. For example, a program that regularly uses competitive sourcing to determine the best value for the taxpayer, invests in IT with clear goals of improving efficiency, etc., could receive a *Yes*. A de-layered management structure that empowers front line managers and that has undergone competitive sourcing (if necessary) would also contribute to a *Yes* answer. For mandatory programs, a *Yes* could require the program to seek policies (e.g., through review of proposals from States) that would reduce unit costs. Also consider if, where possible, there is cross-program and inter-agency coordination on IT issues to avoid redundancies. The program is not required to employ all these strategies to earn a *Yes*. Rather, it should demonstrate that efforts improving efficiency are an established, regular part of program management.

An efficiency measure can be the per-unit cost of outcomes or outputs, a timing target, and other indicator of efficient and productive processes germane to the program. Efficiency measures are likely to be annual measures since they relate to cost.

The answer to this question should describe how measures are used to evaluate the program's success if achieving efficiency and cost effectiveness improvements.

Elements of No: A *No* must be given if the agency and OMB have not reached agreement on efficiency measures that meet PART guidance.

Not Applicable: Not Applicable is not an option for this question.

For more detailed discussion on defining acceptable efficiency measures please see the section called "4. Select Performance Measure" of this document or visit OMB's PART website.[2]

Evidence/Data: Evidence can include efficiency measures, competitive-sourcing plans, IT improvement plans designed to produce tangible productivity and efficiency gains, or IT business cases that document how particular projects improve efficiency.

4.3: Does the program demonstrate improved efficiencies or cost effectiveness in achieving program goals each year?

Purpose: To determine whether management practices have resulted in efficiency gains over the past year.

Elements of Yes: A *Yes* answer needs to clearly explain and provide evidence of each of the following [see Box I-2]:

• The program demonstrated improved efficiency or cost effectiveness over the prior year. When possible, the explanation should include specific information about the program's annual savings over the prior year as well as what the program did to achieve the savings.

---

[2]http://www.omb.gov/part/.

BOX I-2 Question Linkages

If a program received a *No* in Question 3.4, the program must receive a *No* answer to this question.

Efficiency improvements should generally be measured in terms of dollars or time. For example, programs that complete an A-76 competition—an indicator of cost-efficient processes—would contribute to a *Yes* answer, provided that the competition resulted in savings.

Not Applicable: Not Applicable is not an option for this question.

Evidence/Data: Evidence can include meeting performance targets to reduce per unit costs or time, meeting production and schedule targets; or meeting other targets that result in tangible productivity or efficiency gains. Efficiency measures may also be considered in Questions 4.1 and 4.2.

REFERENCES

OMB (Office of Management and Budget). 2006. Program Assessment Rating Tool Guidance 2006-02. Office of Management and Budget. March 2006 [online]. Available: http://www.whitehouse.gov/omb/part/fy2006/2006_guidance_final.pdf [accessed Dec. 17, 2007].